Optimization Software Guide

Frontiers in Applied Mathematics

**Managing Editors
for Practical Computing Series**

W. M. Coughran, Jr.
AT&T Bell Laboratories
Murray Hill, New Jersey

Eric Grosse
AT&T Bell Laboratories
Murray Hill, New Jersey

Frontiers in Applied Mathematics is a series that presents new mathematical or computational approaches to significant scientific problems. Beginning with Volume 4, the series reflects a change in both philosophy and format. Each volume focuses on a broad application of general interest to applied mathematicians as well as engineers and other scientists.

This unique series will advance the development of applied mathematics through the rapid publication of short, inexpensive books that lie on the cutting edge of research.

Frontiers in Applied Mathematics

Vol. 1   Ewing, Richard E., The Mathematics of Reservoir Simulation
Vol. 2   Buckmaster, John D., The Mathematics of Combustion
Vol. 3   McCormick, Stephen F., Multigrid Methods
Vol. 4   Coleman, Thomas F. and Van Loan, Charles, Handbook for Matrix Computations
Vol. 5   Grossman, Robert, Symbolic Computation: *Applications to Scientific Computing*
Vol. 6   McCormick, Stephen F., Multilevel Adaptive Methods for Partial Differential Equations
Vol. 7   Bank, R. E., PLTMG: A Software Package for Solving Elliptic Partial Differential Equations. *Users' Guide 6.0*
Vol. 8   Castillo, José E., Mathematical Aspects of Numerical Grid Generation
Vol. 9   Van Huffel, Sabine and Vandewalle, Joos, The Total Least Squares Problem: *Computational Aspects and Analysis*
Vol. 10 Van Loan, Charles, Computational Frameworks for the Fast Fourier Transform
Vol. 11 Banks, H.T., Control and Estimation in Distributed Parameter Systems
Vol. 12 Cook, L. Pamela, Transonic Aerodynamics: *Problems in Asymptotic Theory*
Vol. 13 Rüde, Ulrich, Mathematical and Computational Techniques for Multilevel Adaptive Methods
Vol. 14 Moré, Jorge J. and Wright, Stephen J., Optimization Software Guide

Optimization Software Guide

Jorge J. Moré
Stephen J. Wright
Argonne National Laboratory

Society for Industrial and Applied Mathematics
siam.
Philadelphia 1993

The book has been authored by a contractor of the United States Government under contract W-31-109-Eng-38. Accordingly, the United States Government retains a nonexclusive royalty-free license to publish or reproduce the published form of this contribution, or allow others to do so for the United States Government.

Library of Congress Cataloging-in-Publication Data

Moré, Jorge J.
    Optimization software guide   /   Jorge J. Moré and Stephen J. Wright.
      p.   cm.   —   (Frontiers in applied mathematics   :   vol. 14)
    Included bibliographical references.
    ISBN   0-89871-322-6
    1. Mathematical optimization—Data processing.   2. Numerical calculation—Data processing.   3. Computer software.   I. Wright, Steve (Steve J.)   II. Title.   III. Series: Frontiers in applied mathematics   :   14.
    QA402.5.M67   1993
    519.3'0285'53—dc20   93-33771

All rights reserved. Printed in the United States of America. No part of this book may be reproduced, stored, or transmitted in any manner without the written permission of the Publisher. For information, write the Society for Industrial and Applied Mathematics, 3600 University City Science Center, Philadelphia, Pennsylvania 19104-2688.

Copyright © 1993 by the Society for Industrial and Applied Mathematics.
Second Printing 1994.

siam. is a registered trademark.

To Our Parents

Contents

Preface ... xi

I OVERVIEW OF ALGORITHMS ... 1

CHAPTER 1. Optimization Problems and Software ... 3

CHAPTER 2. Unconstrained Optimization ... 7

CHAPTER 3. Nonlinear Least Squares ... 15

CHAPTER 4. Nonlinear Equations ... 21

CHAPTER 5. Linear Programming ... 27

CHAPTER 6. Quadratic Programming ... 35

CHAPTER 7. Bound-Constrained Optimization ... 39

CHAPTER 8. Constrained Optimization ... 45

CHAPTER 9. Network Optimization ... 53

CHAPTER 10. Integer Programming ... 59

CHAPTER 11. Miscellaneous Optimization Problems ... 63

II SOFTWARE PACKAGES ... 65

Software Classification ... 67
    AMPL ... 69
    BQPD ... 69
    BT ... 70

BTN	71
CNM	73
CONOPT	73
CONSOL-OPTCAD	74
CPLEX	76
C-WHIZ	78
DFNLP	78
DOC	79
DOT	80
FortLP	81
FSQP	82
GAMS	83
GAUSS	84
GENESIS	85
GENOS	87
GINO	87
GRG2	88
HOMPACK	89
IMSL Fortran and C Library	90
LAMPS	93
LANCELOT	94
LBFGS	95
LINDO	96
LINGO	97
LNOS	98
LPsolver	100
LSGRG2	100
LSNNO	101
LSSOL	102
M1QN2 and M1QN3	104
MATLAB	105
MINOS	106
MINPACK-1	108
MIPIII	109
MODULOPT	109
NAG C library	110
NAG Fortran Library	111
NETFLOW	112
NETSOLVE	113
NITSOL	114
NLPE	115
NLPQL	116
NLPQLB	117
NLSFIT	118

NLSSOL	119
NLPSPR	121
NPSOL	122
OB1	123
ODRPACK	124
OPSYC	126
OptiA	126
OPTIMA Library	128
OPTPACK	129
OSL	130
PC-PROG	131
PITCON	132
PORT 3	133
PROC NLP	134
Q01SUBS	136
QAPP	136
QPOPT	137
SPEAKEASY	139
SQP	140
TENMIN	140
TENSOLVE	142
TN/TNBC	143
TNPACK	143
UNCMIN	144
VE08	146
VE10	147
VIG and VIMDA	148
What's *Best!*	148

Appendix: Internet Software   151

References   153

Preface

This book presents information on the current state of numerical optimization software that we collected in preparation for a short course on *Numerical Optimization Algorithms and Software*. This information was presented at the SIAM Optimization meeting in May 1992 and at the SIAM Annual meeting in July 1992.

Part I contains an overview of algorithms for different classes of optimization problems. The treatment here is by no means exhaustive. We aim instead to give the reader a feeling for each of the algorithms implemented in the software of Part II, along with pointers to books and survey papers in which more thorough and rigorous discussions can be found. We hope that Part I includes enough information to allow the reader to identify software packages worthy of further investigation.

Part II comprises product descriptions of about a page each, provided by software vendors and individual researchers. Although we made an effort to contact as wide and diverse a collection of vendors as possible, our survey of optimization software is by no means complete. We hope that we have listed the most widely used packages within each category, along with some newer or less well known entries. We plan to expand and update the listings as further information becomes available.

We have refrained from making comparisons of the software on the basis of performance. This task is clearly impossible for the large number of packages included in this guide. Even if a standard test set were available—which is not the case for most problem classes—the continuing evolution of most software packages would make any comparison obsolete within months.

We have also omitted information on pricing of the software. Prices tend to change even more rapidly than the software itself and vary according to whether the buyer is in academia or industry.

This book does not mention software for such widely studied classes of problem as stochastic programming, optimal control, and parameter estimation. This is because no simple, catch-all mathematical formulation exists for each of these classes that can serve as a basis for efficient algorithm design. For example, to devise an efficient algorithm for an optimal control problem, we must know whether the dynamics of the underlying system are governed by ordinary or partial differential equations, and whether these dynamics are stochastic or deterministic. In stochastic programming, the choice of algorithm may differ according to whether the problem is two-stage or multistage, and according to whether the cost functions are linear or nonlinear. Certainly, problems such as these can be formulated as general optimization problems, but the price paid for ignoring the special structure is a significant (often ruinous) loss of efficiency.

Acknowledgments

This book would not have been possible without the support of the many contributors who supplied the detailed information on software packages that makes up Part II. In many cases, the descriptions were prepared by the contributors themselves and appear without modification.

We are also grateful to the support provided by the Office of Scientific Computing of the U.S. Department of Energy. Their support of research on algorithms and software for optimization is reflected in many of the contributions in this book. We also thank Gail Pieper for the numerous suggestions for improvement; her advice was invaluable. Finally, Susan Ciambrano, our editor at SIAM, deserves special thanks for her patience and encouragement.

Part I
OVERVIEW OF ALGORITHMS

Chapter 1
Optimization Problems and Software

Optimization problems occur in all areas of science and engineering, arising whenever there is a need to minimize (or maximize) an *objective function* that depends on a set of *variables* while satisfying some *constraints*. Before an optimization problem can be solved by the software discussed in this book, it must be formulated as one of the standard optimization paradigms. This task often requires interaction between the user and either a user-friendly software interface or a user-friendly optimization consultant. To help in this process, we sketch in this chapter the paradigms for which general-purpose software is available. Additional information on each problem class, and the corresponding algorithms and software, can be found in the remainder of Part I.

We use notation that is standard in the optimization literature. Lowercase letters denote vectors with real components or functions defined on such vectors. For example, $x \in \mathbb{R}^n$ is a real vector with n components, while $f : \mathbb{R}^n \to \mathbb{R}^m$ is a vector-valued function. Uppercase letters denote matrices of real numbers; for example, $A \in \mathbb{R}^{m \times n}$ is an $m \times n$ real matrix. Subscripts on vectors are used both for components (x_i is the ith component of x) and iteration indices (x_k is the kth member of a sequence of iterates). The intention should be clear from the context.

Optimization problems may be formally specified by defining a vector of variables $x \in \mathbb{R}^n$, an objective function $f_0 : \mathbb{R}^n \to \mathbb{R}$, and a constraint function $c : \mathbb{R}^n \to \mathbb{R}^p$. Some of the components of c may be *inequality constraints* of the form $c_i(x) \leq 0$ for each i in some index set $\mathcal{I}$, while other components may represent *equality constraints* of the form $c_i(x) = 0$ for each i in an index set $\mathcal{E}$. There may also be two-sided constraints of the form $l_i \leq c_i(x) \leq u_i$, where l_i and u_i are lower and upper bounds; the special cases where $c_i(x) = x_i$ are *bound constraints*.

In the nonlinearly constrained optimization (or *nonlinear programming*) problem, f_0 and c are general nonlinear functions of x. Two common (and equivalent) formulations of this problem are

$$\min \{f_0(x) : c_i(x) \leq 0,\ i \in \mathcal{I},\ \ c_i(x) = 0,\ i \in \mathcal{E}\}$$

and

$$\min \{f_0(x) : c(x) = 0,\ l \leq x \leq u\}.$$

Different software packages assume different formulations, but these two are typical, and both are used during our discussion of nonlinear programming algorithms.

In *linear programming* problems, the objective function and all constraints are linear. The standard form for such problems is

$$\min \left\{ c^T x : Ax = b, \ x \geq 0 \right\},$$

where $c \in \mathbb{R}^n$ is a cost vector and $A \in \mathbb{R}^{m \times n}$ is a constraint matrix. Software for linear programming usually allows more convenient formulations to be used, such as

$$\min \left\{ c^T x : l_i \leq a_i^T x \leq u_i, \ i \in \mathcal{I}, \ l_i \leq x_i \leq u_i, \ i \in \mathcal{B} \right\}.$$

In this formulation equality constraints are enforced by setting $l_i = u_i$ for $i \in \mathcal{I}$.

Quadratic programming problems have linear constraints and quadratic objective functions, that is,

$$\min \left\{ c^T x + \tfrac{1}{2} x^T Q x : a_i^T x \leq b_i, \ i \in \mathcal{I}, \ a_i^T x = b_i, \ i \in \mathcal{E} \right\},$$

where $Q \in \mathbb{R}^{n \times n}$ is a symmetric matrix. These problems are convex if Q is positive semidefinite, and nonconvex otherwise. Some convex quadratic programming problems can be formulated more naturally as the *constrained linear least squares* problem

$$\min \left\{ \tfrac{1}{2} \|Cx - d\|_2^2 : a_i^T x \leq b_i, \ i \in \mathcal{I}, \ a_i^T x = b_i, \ i \in \mathcal{E} \right\},$$

where the coefficient matrix C is not necessarily square.

Bound-constrained optimization problems have the form

$$\min \left\{ f_0(x) : l \leq x \leq u \right\}.$$

The only constraints are simple bounds on the components of x. When neither bounds nor general constraints are present, we have an *unconstrained* optimization problem.

In *nonlinear least squares* problems, the objective function has the special form

$$f_0(x) = \tfrac{1}{2} \sum_{i=1}^{m} f_i(x)^2,$$

where each component function f_i is often called a residual. We can also write f_0 as

$$f_0(x) = \tfrac{1}{2} \|f(x)\|_2^2$$

by grouping the residuals into a vector-valued function $f : \mathbb{R}^n \to \mathbb{R}^m$.

The solution to a *system of nonlinear equations* specified by the mapping $f : \mathbb{R}^n \to \mathbb{R}^n$ is a vector x such that $f(x) = 0$. Some algorithms for solving systems of nonlinear equations are closely related to algorithms for unconstrained optimization and nonlinear least squares, since they aim to solve

$$\min \{\|f(x)\| : x \in \mathbb{R}^n\},$$

where the norm $\|\cdot\|$ is usually the l_2 norm on $\mathbb{R}^n$.

In *network optimization* problems the functions f_0 and c_i have a special structure that arises from a graph consisting of arcs and nodes. The constraints are usually linear, and each constraint c_i usually involves only one or two components of x. The function f_0 may be either linear or nonlinear, but it is usually *separable*; that is,

$$f_0(x) = \sum_{i=1}^{n} f_i(x_i),$$

where each f_i is a scalar function of the scalar arguments x_i.

In the preceding discussion we assumed that all unknowns are real numbers, that is, $x \in \mathbb{R}^n$. When the components of x are integers, we have an *integer-programming* problem. The term *mixed-integer programming* is used when x is a combination of real and integer components. The allowable values of x may be restricted further, for example, to the set $\{0, 1\}$. Integer-programming problems are usually more difficult to solve than problems in which all components of x are real, and production software exists only for integer linear programming and some special cases of integer quadratic programming.

Most optimization problems cannot be classified uniquely according to the taxonomy above. For example, a linear programming problem is also a quadratic programming problem, which is also a nonlinear programming problem. These ambiguities should be resolved by placing such problems in the most restricted class for which they qualify. This strategy increases the chance that software will be able to take full advantage of the problem structure.

The discussions of each problem class in the remainder of Part I include pointers to the relevant software descriptions in Part II. These descriptions are arranged alphabetically by software package name. However, a cross-reference listing, grouped according to our taxonomy of problem classes, appears at the start of Part II.

Chapter 2
Unconstrained Optimization

The unconstrained optimization problem is central to the development of optimization software since constrained optimization algorithms are often extensions of unconstrained algorithms, while nonlinear least squares and nonlinear equation algorithms tend to be specializations. In the unconstrained optimization problem
$$\min\{f(x) : x \in \mathbb{R}^n\},$$
we seek a local minimizer of a real-valued function f defined on $\mathbb{R}^n$, that is, a vector $x^* \in \mathbb{R}^n$ such that $f(x^*) \leq f(x)$ for all $x \in \mathbb{R}^n$ near x^*. We do not discuss global minimization algorithms because at present there is no widely available code for global minimization.

Newton's method gives rise to a wide and important class of algorithms that require computation of the gradient vector,
$$\nabla f(x) = \begin{pmatrix} \partial_1 f(x) \\ \vdots \\ \partial_n f(x) \end{pmatrix},$$
and the Hessian matrix,
$$\nabla^2 f(x) = \left(\partial_j \partial_i f(x)\right).$$
Although the computation or approximation of the Hessian matrix can be a time-consuming operation, there are many problems for which this computation is justified. We describe algorithms that assume $\nabla^2 f(x)$ to be available before moving on to a discussion of algorithms that do not make this assumption.

The basic Newton method chooses the iterates by minimizing the quadratic model
$$q_k(s) = f(x_k) + \nabla f(x_k)^T s + \tfrac{1}{2} s^T \nabla^2 f(x_k) s$$
of the function about the current iterate x_k. When the Hessian matrix $\nabla^2 f(x_k)$ is positive definite, q_k has a unique minimizer that can be obtained by solving the symmetric $n \times n$ linear system
$$\nabla^2 f(x_k) s_k = -\nabla f(x_k).$$

The next iterate is then $x_{k+1} = x_k + s_k$. Convergence is guaranteed if x_0 is sufficiently close to a local minimizer x^* at which $\nabla^2 f(x^*)$ is positive definite. Moreover, the rate of convergence is quadratic, that is,

$$\|x_{k+1} - x^*\| \leq \beta \|x_k - x^*\|^2,$$

for some positive constant β. If these conditions are not satisfied, however, the basic Newton method may need to be modified in order to achieve convergence.

Versions of Newton's method are implemented in BTN, GAUSS, IMSL, LANCELOT, NAG, OPTIMA, PORT 3, PROC NLP, TENMIN, TN, TNPACK, UNCMIN, and VE08. These codes enforce convergence when the initial point x_0 is not close to a minimizer x^* by using either a *line-search* or a *trust-region* approach. As we shall see, these two approaches differ mainly in the way they treat indefiniteness in the Hessian matrix, $\nabla^2 f(x_k)$.

Line-search methods generate the iterates by setting $x_{k+1} = x_k + \alpha_k d_k$, where d_k is a search direction and $\alpha_k > 0$ is chosen so that $f(x_{k+1}) < f(x_k)$. Most line-search versions of the basic Newton method generate the direction d_k by modifying the Hessian matrix $\nabla^2 f(x_k)$ to ensure that the quadratic model q_k of the function has a unique minimizer. The *modified Cholesky* decomposition approach adds positive quantities to the diagonal of $\nabla^2 f(x_k)$ during the Cholesky factorization. As a result, a diagonal matrix E_k with nonnegative diagonal entries is generated such that $\nabla^2 f(x_k) + E_k$ is positive definite. Given this decomposition, the search direction d_k is obtained by solving

$$\left(\nabla^2 f(x_k) + E_k\right) d_k = -\nabla f(x_k).$$

After d_k is found, a line-search procedure is used to choose an $\alpha_k > 0$ that approximately minimizes f along the ray $\{x_k + \alpha d_k : \alpha > 0\}$.

The Newton codes in GAUSS, NAG, and OPTIMA use line-search methods. The algorithms used in these codes for determining α_k rely on quadratic or cubic interpolation of the univariate function $\phi(\alpha) = f(x_k + \alpha d_k)$ in their search for a suitable α_k. An elegant and practical criterion for a suitable α_k is to require α_k to satisfy the *sufficient decrease* condition

$$f(x_k + \alpha_k d_k) \leq f(x_k) + \mu \alpha_k \nabla f(x_k)^T d_k$$

and the *curvature* condition

$$\left|\nabla f(x_k + \alpha_k d_k)^T d_k\right| \leq \eta \left|\nabla f(x_k)^T d_k\right|,$$

where μ and η are two constants with $0 < \mu < \eta < 1$. The sufficient decrease condition guarantees, in particular, that $f(x_{k+1}) < f(x_k)$, while the curvature condition requires that α_k be not too far from a minimizer of ϕ. Requiring an accurate minimizer is generally wasteful of function and gradient evaluations, so codes typically use $\mu = 0.001$ and $\eta = 0.9$ in these conditions.

The trust-region approach can be motivated by noting that the quadratic model q_k is a useful model of the function f only near x_k. When the Hessian matrix $\nabla^2 f(x_k)$ is indefinite, the quadratic q_k is unbounded below, so q_k is obviously a poor model of $f(x_k + s)$ when s is large. Therefore, it is reasonable to select the step s_k by solving the subproblem

$$\min\{q_k(s) : \|D_k s\|_2 \leq \Delta_k\}$$

for some $\Delta_k > 0$ and scaling matrix D_k. The trust-region parameter Δ_k is adjusted between iterations according to the agreement between predicted and actual reduction in the function f as measured by the ratio

$$\rho_k = \frac{f(x_k) - f(x_k + s_k)}{f(x_k) - q_k(s_k)}.$$

If there is good agreement ($\rho_k \approx 1$), then Δ_k is increased; if the agreement is poor (ρ_k small or ρ_k negative), then Δ_k is decreased. The decision to accept the step s_k is also based on ρ_k; usually $x_{k+1} = x_k + s_k$ if $\rho_k \geq \sigma_0$, where σ_0 is small (typically, 10^{-4}); otherwise $x_{k+1} = x_k$.

The Newton codes in IMSL, LANCELOT, PORT 3, and PROC NLP use trust-region methods with different algorithms for choosing the step s_k. Most of these algorithms rely on the observation that there is a $\lambda_k \geq 0$ such that

$$(2.1) \qquad \left(\nabla^2 f(x_k) + \lambda_k D_k^T D_k\right) s_k = -\nabla f(x_k),$$

where either $\lambda_k = 0$ and $\|D_k s_k\|_2 \leq \Delta_k$, or $\lambda_k > 0$ and $\|D_k s_k\|_2 = \Delta_k$. An appropriate λ_k is determined by an iterative process in which (2.1) is solved for each trial value of λ_k.

The algorithm implemented in the TENMIN package extends Newton's method by forming low-rank approximations to the third- and fourth-order terms in the Taylor series approximation. These approximations are formed by using matching conditions that require storage of the gradient (but not the Hessian) on a number of previous iterations.

The line-search and trust-region techniques that we have described are suitable if the number of variables n is not too large, because the cost per iteration is of order n^3. Codes for problems with a large number of variables tend to use iterative techniques for obtaining a direction d_k in a line-search method or a step s_k in a trust-region method. These codes are usually called *truncated Newton* methods because the iterative technique is stopped (truncated) as soon as a termination criterion is satisfied. For example, the codes in BTN, TN, TNPACK, and VE08 use a line-search method in which the direction d_k satisfies

$$\|\nabla^2 f(x_k) d_k + \nabla f(x_k)\| \leq \eta_k \|\nabla f(x_k)\|$$

for some $\eta_k \in (0, 1)$; the LANCELOT codes use a similar idea in the context of trust-region methods. Conjugate gradient algorithms mesh well with truncated Newton methods because of their desirable numerical properties.

Preconditioning is necessary in order to improve the efficiency and reliability of the conjugate gradient method; effective preconditioners include those based on the incomplete Cholesky factorization and symmetric successive overrelaxation.

So far, we have assumed that the Hessian matrix is available, but the algorithms are unchanged if the Hessian matrix is replaced by a reasonably accurate approximation. The most common method for obtaining such an approximation is to use differences of gradient values. If forward differences are used, then the ith column of the Hessian matrix is replaced by

$$\frac{\nabla f(x_k + h_i e_i) - \nabla f(x_k)}{h_i}$$

for some suitable choice of difference parameter h_i, where e_i is the vector with 1 in the ith position and zeros elsewhere. Similarly, if central differences are used, the ith column is replaced by

$$\frac{\nabla f(x_k + h_i e_i) - \nabla f(x_k - h_i e_i)}{2h_i}.$$

An appropriate choice of difference parameter h_i can be difficult. Rounding errors overwhelm the calculation if h_i is too small, while truncation errors dominate if h_i is too large. Newton codes rely on forward differences, since they often yield sufficient accuracy for reasonable values of h_i; central differences are more accurate but they require twice the work ($2n$ gradient evaluations against n evaluations).

Variants of Newton's method for problems with a large number of variables cannot use the above techniques to approximate the Hessian matrix because the cost of n gradient evaluations is prohibitive. For problems with a sparse Hessian matrix, however, it is possible to use specialized techniques based on graph coloring that allow difference approximations to the Hessian matrix to be computed in many fewer than n gradient evaluations. For example, if the Hessian matrix has bandwidth $2b + 1$, then only $b + 1$ gradient evaluations are required.

VE08 is designed to solve optimization problems with a large number of variables where f is a *partially separable* function, that is, f can be written in the form

$$f(x) = \sum_{i=1}^{m} f_i(x),$$

where each function $f_i : \mathbb{R}^n \to \mathbb{R}$ has an invariant subspace

$$\mathcal{N}_i = \{w \in \mathbb{R}^n : f_i(x + w) = f_i(x) \text{ for all } x \in \mathbb{R}^n\}$$

whose dimension is large relative to the number of variables n. This is the case, in particular, if f_i depends only on a small number (typically, fewer than ten) of the components of x. Functions with sparse Hessian matrices are partially separable; indeed, most functions that arise in large-scale problems are partially separable. An advantage of algorithms designed for these problems is that techniques for approximating a dense Hessian matrix (for example, forward differences) can be used to approximate the nontrivial part of the element Hessian matrix $\nabla^2 f_i(x)$. Approximations to $\nabla^2 f(x)$ can be obtained by summing the approximations to $\nabla^2 f_i(x)$.

Quasi-Newton or *variable metric* methods can be used when the Hessian matrix is difficult or time-consuming to evaluate. Instead of obtaining an estimate of the Hessian matrix at a single point, these methods gradually build up an approximate Hessian matrix by using gradient information from some or all of the previous iterates x_k visited by the algorithm. Given the current iterate x_k, and the approximate Hessian matrix B_k at x_k, the linear system

$$B_k d_k = -\nabla f(x_k)$$

is solved to generate a direction d_k. The next iterate is then found by performing a line search along d_k and setting $x_{k+1} = x_k + \alpha_k d_k$. The question is then: How can we use the function and gradient information from points x_k and x_{k+1} to improve the quality of the approximate Hessian matrix B_k? In other words, how do we obtain the new approximate Hessian matrix B_{k+1} from the previous approximation B_k?

The key to this question depends on what is sometimes called the fundamental theorem of integral calculus. If we define

$$s_k = x_{k+1} - x_k, \qquad y_k = \nabla f(x_{k+1}) - \nabla f(x_k),$$

then this theorem implies that

$$\left\{ \int_0^1 \nabla^2 f(x_k + t s_k) \, dt \right\} s_k = y_k.$$

The matrix in braces can be interpreted as the average of the Hessian matrix on the line segment $[x_k, x_k + s_k]$; this result states that when this matrix is applied to s_k, the result is y_k. In view of these observations, we can make B_{k+1} mimic the behavior of $\nabla^2 f$ by enforcing the *quasi-Newton condition*

$$B_{k+1} s_k = y_k.$$

This condition can be satisfied by making a simple low-rank update to B_k. The most commonly used family of updates is the Broyden class of rank-two updates, which have the form

$$B_{k+1} = B_k - \frac{B_k s_k (B_k s_k)^T}{s_k^T B_k s_k} + \frac{y_k y_k^T}{y_k^T s_k} + \phi_k \left[s_k^T B_k s_k \right] v_k v_k^T,$$

where $\phi_k \in [0, 1]$ and

$$v_k = \left[\frac{y_k}{y_k^T s_k} - \frac{B_k s_k}{s_k^T B_k s_k} \right].$$

BFGS

The choice $\phi_k = 0$ gives the Broyden–Fletcher–Goldfarb–Shanno update, which practical experience and some theoretical analysis has shown to be the method of choice in most circumstances. The Davidon–Fletcher–Powell update, which was proposed earlier, is obtained by setting $\phi_k = 1$. These two update formulae are known universally by their initials BFGS and DFP, respectively.

Updates in the Broyden class remain positive definite as long as $y_k^T s_k > 0$. Although the condition $y_k^T s_k > 0$ holds automatically if f is strictly convex, this condition can be enforced for all functions by requiring that α_k satisfy the curvature condition. Some codes avoid enforcing the curvature condition by skipping the update if $y_k^T s_k \leq 0$.

GAUSS, IMSL, MATLAB, NAG, OPTIMA, and PROC NLP implement quasi-Newton methods. These codes differ in the choice of update (usually BFGS), line-search procedure, and the way in which B_k is stored and updated. We can update B_k by either updating the Cholesky decomposition of B_k or by updating the inverse of B_k. In either case, the cost of updating the search direction by solving the system $B_k d_k = -\nabla f(x_k)$ is on the order of n^2 operations. Updating the Cholesky factors is widely regarded as more reliable, while updating the inverse of B_k is less complicated. Indeed, if we define $H_k = B_k^{-1}$ then a BFGS update of B_k is equivalent to the following update of H_k:

$$H_{k+1} = \left(I - \frac{s_k y_k^T}{y_k^T s_k}\right) H_k \left(I - \frac{y_k s_k^T}{y_k^T s_k}\right) + \frac{s_k s_k^T}{y_k^T s_k}.$$

When we store H_k explicitly, the direction d_k is obtained from the matrix-vector product $d_k = -H_k \nabla f(x_k)$.

The availability of quasi-Newton methods renders steepest-descent methods obsolete. Both types of algorithms require only first derivatives, and both require a line search. The quasi-Newton algorithms require slightly more operations to calculate an iterate and somewhat more storage, but in almost all cases, these additional costs are outweighed by the advantage of superior convergence.

At first glance quasi-Newton methods may seem unsuitable for large problems because the approximate Hessian matrices and inverse Hessian matrices are generally dense. This is not the case, as explicit storage of B_k or H_k as $n \times n$ matrices is not necessary. For example, the above expression for the BFGS update of H_k makes it clear that we can compute $H_k \nabla f(x_k)$ if we know the initial matrix H_0; the subsequent update vectors s_i, y_i; and their inner products $y_i^T s_i$ for $0 \leq i < k$. If H_0 is chosen to be a diagonal matrix, the necessary information can be stored in about $2nk$ words of memory. *Limited-memory quasi-Newton* methods make use of these ideas to cut down on storage for large problems. They store only the s_i and y_i vectors from the previous few iterates (typically, five) and compute the matrix-vector $H_k \nabla f(x_k)$ by a recursion that requires roughly $16\,nm$ operations. The LBFGS code is an implementation of the limited-memory BFGS algorithms. The codes M1QN2 and M1QN3 are the same as LBFGS, except that they allow the user to specify a preconditioning technique.

Nonlinear conjugate gradient algorithms make up another popular class of algorithm for large-scale optimization. These algorithms can be derived as extensions of the conjugate gradient algorithm or as specializations of limited-memory quasi-Newton methods. Given an iterate x_k and a direction d_k, a line search is performed along d_k to produce $x_{k+1} = x_k + \alpha_k d_k$. The Fletcher–Reeves variant of the nonlinear conjugate algorithm generates d_{k+1} from the simple recursion

$$d_{k+1} = -\nabla f(x_{k+1}) + \beta_k d_k, \qquad \beta_k = \left(\frac{\|\nabla f(x_{k+1})\|_2}{\|\nabla f(x_k)\|_2} \right)^2.$$

The method's performance is sometimes enhanced by *re-starting*, that is, periodically setting β_k to zero. The Polak–Ribiere variant of conjugate gradient defines β_k as

$$\beta_k = \frac{[\nabla f(x_{k+1}) - \nabla f(x_k)]^T \nabla f(x_{k+1})}{\nabla f(x_k)^T \nabla f(x_k)}.$$

These two definitions of β_k are equivalent when f is quadratic, but not otherwise. Numerical testing suggests that the Polak–Ribiere method tends to be more efficient than the Fletcher–Reeves method. IMSL, NAG, OPTIMA, OPTPACK, and PROC NLP contain nonlinear conjugate gradient codes.

An algorithm that may be appropriate when the gradient of f is hard to calculate, or when the function value contains noise, is the *nonlinear simplex* method. For an n-dimensional problem, this method maintains a simplex of $n+1$ points (a triangle in two dimensions, or a pyramid in three dimensions). The simplex moves, expands, contracts, and distorts its shape as it attempts to find a minimizer. This method is slow and can be applied only to problems in which n is small. It is, however, extremely popular, since it requires the user to supply only function values, not derivatives. IMSL, MATLAB, NAG, PORT 3, and PROC NLP contain implementations of this method.

Notes and References

Unconstrained optimization algorithms are discussed in the books of Dennis and Schnabel [13]; Fletcher [15]; and Gill, Murray, and Wright [18]. These books cover line-search and trust-region methods; note that Fletcher uses the term *restricted-step* methods instead of trust-region methods. Additional information can be obtained from the survey papers by Moré and Sorensen [27], Dennis and Schnabel [14], and Nocedal [31]. We did not discuss global minimization algorithms because at present there is no widely available code for global minimization. However, this is an active research area, to which the *Journal on Global Optimization* is exclusively dedicated. Rinnooy Kan and Timmer [33] survey the major global optimization algorithms.

Chapter 3
Nonlinear Least Squares

The nonlinear least squares problem has the general form

$$\min\{r(x) : x \in \mathbb{R}^n\},$$

where r is the function defined by

$$r(x) = \tfrac{1}{2}\|f(x)\|_2^2$$

for some vector-valued function f that maps $\mathbb{R}^n$ to $\mathbb{R}^m$.

Least squares problems often arise in data-fitting applications. Suppose that some physical or economic process is modeled by a nonlinear function ϕ that depends on a parameter vector x and time t. If b_i is the actual output of the system at time t_i, then the residual

$$\phi(x, t_i) - b_i$$

measures the discrepancy between the predicted and observed outputs of the system at time t_i. A reasonable estimate for the parameter x may be obtained by defining the ith component of f by $f_i(x) = \phi(x, t_i) - b_i$, and solving the least squares problem with this definition of f.

From an algorithmic point of view, the feature that distinguishes least squares problems from the general unconstrained optimization problem is the structure of the Hessian matrix $\nabla^2 r(x)$. The Jacobian matrix of f,

$$f'(x) = \big(\partial_1 f(x), \ldots, \partial_n f(x)\big),$$

can be used to express the gradient of r since

$$\nabla r(x) = f'(x)^T f(x).$$

Similarly, $f'(x)$ is part of the Hessian matrix $\nabla^2 r(x)$ since

$$\nabla^2 r(x) = f'(x)^T f'(x) + \sum_{i=1}^{m} f_i(x) \nabla^2 f_i(x).$$

To calculate the gradient of r, we need to calculate the Jacobian matrix $f'(x)$. Having done so, we know the first term in the Hessian matrix $\nabla^2 r(x)$ without doing any further evaluations. Nonlinear least squares algorithms exploit this structure.

In many practical circumstances, the first term $f'(x)^T f'(x)$ in $\nabla^2 r(x)$ is more important than the second term, most notably when the residuals $f_i(x)$ are small at the solution. Specifically, we say that a problem has small residuals if, for all x near a solution, the quantities

$$|f_i(x)| \, \|\nabla^2 f_i(x)\|, \qquad i = 1, 2, \ldots, n$$

are small relative to the smallest eigenvalue of $f'(x)^T f'(x)$.

An algorithm that is particularly suited to the small-residual case is the *Gauss–Newton* algorithm, in which the Hessian $\nabla^2 r$ is approximated by its first term. In a line-search version of the Gauss–Newton algorithm, the search direction d_k from the current iterate x_k satisfies the linear system

(3.1) $$\left(f'(x_k)^T f'(x_k) \right) d_k = -f'(x_k)^T f(x_k).$$

Any solution d_k of (3.1) is a descent direction for r, since

$$d_k^T \nabla r(x_k) = -\|f'(x_k) d_k\|_2^2 < 0,$$

unless $\nabla r(x_k) = 0$. Gauss–Newton codes perform a line search along the direction d_k to obtain the new iterate $x_{k+1} = x_k + \alpha_k d_k$. The suitability of a candidate α_k can be determined, as in the case of unconstrained minimization, by enforcing the sufficient decrease condition and the curvature condition.

When $f'(x_k)$ has rank less than n, the system (3.1) has a multiplicity of solutions. In these circumstances, Gauss–Newton algorithms choose a particular solution. For example, the choice

$$d_k = -f'(x_k)^+ f(x_k),$$

where A^+ denotes the Moore–Penrose pseudo-inverse of A, corresponds to the solution of least l_2 norm. Since this choice is expensive to compute and requires the determination of the rank of $f'(x_k)$, codes tend to obtain a direction d_k that satisfies (3.1) by using less expensive techniques.

The Gauss–Newton algorithm is used, usually with enhancements, in much of the software for nonlinear least squares. It is a component of the algorithms used by DFNLP, MATLAB, NAG, OPTIMA, and TENSOLVE. In many of these codes the Gauss–Newton model is augmented by a matrix S_k; as a consequence the direction d_k satisfies

$$\left(f'(x_k)^T f'(x_k) + S_k \right) d_k = -f'(x_k)^T f(x_k).$$

The purpose of S_k is to guarantee fast local convergence while retaining the global convergence properties of the Gauss–Newton method.

The NAG routines use a Gauss–Newton search direction whenever a sufficiently large decrease in r is obtained at the previous iteration. Otherwise, second-derivative information is obtained from user-supplied function evaluation routines, quasi-Newton approximations, or difference approximations. Using this information, the software attempts to find a more accurate approximation to the Newton direction than the Gauss–Newton direction is able to provide.

The TENSOLVE software augments the Gauss–Newton model with a low-rank tensor approximation to the second-order term. It has been observed to converge faster than standard Gauss–Newton on many problems, particularly when the Jacobian matrix is rank deficient at the solution.

The *Levenberg–Marquardt* algorithm can be thought of as a trust-region modification of the Gauss–Newton algorithm. Levenberg–Marquardt steps s_k are obtained by solving subproblems of the form

$$(3.2) \qquad \min\left\{ \tfrac{1}{2} \|f'(x_k)s + f(x_k)\|_2^2 : \|D_k s\|_2 \leq \Delta_k \right\},$$

for some $\Delta_k > 0$ and scaling matrix D_k. The trust-region radius Δ_k is adjusted between iterations according to the agreement between predicted and actual reduction in the objective function r. For a step s_k to be accepted, the ratio

$$\rho_k = \frac{r(x_k) - r(x_k + s_k)}{r(x_k) - \tfrac{1}{2}\|f'(x_k)s_k + f(x_k)\|_2^2}$$

must exceed a small positive number σ_0. (A typical value is $\sigma_0 = 10^{-4}$.) If s_k fails this test, then Δ_k is decreased and s_k is recalculated. When ρ_k is close to 1, the trust region for the next iteration is expanded by choosing the new radius to be larger than Δ_k.

Levenberg–Marquardt codes usually determine s_k by noting that the solution s_k of (3.2) also satisfies the equation

$$(3.3) \qquad \left(f'(x_k)^T f'(x_k) + \lambda_k D_k^T D_k \right) s_k = -f'(x_k)^T f(x_k),$$

for some $\lambda_k \geq 0$. The Lagrange multiplier λ_k is zero if the minimum-norm Gauss–Newton step is smaller than Δ_k; otherwise λ_k is chosen so that $\|D_k s_k\|_2 = \Delta_k$. Equations (3.3) are simply the normal equations for the least squares problem

$$(3.4) \qquad \min\left\{ \left\| \begin{bmatrix} f'(x_k) \\ \lambda_k^{1/2} D_k \end{bmatrix} s + \begin{bmatrix} f(x_k) \\ 0 \end{bmatrix} \right\|_2^2 : s \in \mathbb{R}^n \right\}.$$

Efficient factorization of the coefficient matrix in (3.4) can be performed by a combination of Householder and Givens transformations.

The Levenberg–Marquardt algorithm has proved to be an effective and popular way to solve nonlinear least squares problems. MINPACK-1 contains

Levenberg–Marquardt codes in which the Jacobian matrix may be either supplied by the user or calculated by using finite differences. IMSL, MATLAB, ODRPACK, and PROC NLP also contain Levenberg–Marquardt routines.

The algorithms in ODRPACK solve unconstrained nonlinear least squares problems and orthogonal distance regression problems, including those with implicit models and multiresponse data. The simplest case of an orthogonal distance regression problem occurs when we are given observations y_i at times t_i and a model $\phi(t_i, \cdot)$ for each y_i at time t_i. In this case, we aim to find sets of parameters $x \in \mathbb{R}^n$ and corrections δ_i that solve the minimization problem

$$\min \left\{ \sum_{i=1}^{m} [\phi(t_i + \delta_i; x) - y_i]^2 + \delta_i^2 : x \in \mathbb{R}^n,\ \delta_i \in \mathbb{R} \right\}.$$

The ODRPACK algorithms use a trust-region Levenberg–Marquardt method that exploits the structure of this problem, so that there is little difference between the cost per iteration for this problem and the standard least squares problem in which the δ_i are fixed at zero.

The Gauss–Newton and Levenberg–Marquardt algorithms usually exhibit quadratic convergence for zero-residual ($r(x^*) = 0$) problems. Otherwise, the convergence is only linear. It would seem that something could be gained by treating a nonlinear least squares problem as a general unconstrained minimization problem and applying quasi-Newton algorithms to it, since quasi-Newton algorithms are superlinearly convergent. A simple hybrid strategy, implemented in the PROC NLP algorithms, combines the Gauss–Newton and BFGS quasi-Newton algorithms. In this approach, the Gauss–Newton step is used if it decreases the value of r by a factor of 5. If not, the BFGS step is taken. The initial approximate Hessian for BFGS is taken to be the initial Gauss–Newton matrix $f'(x_0)^T f'(x_0)$. For zero-residual problems, Gauss–Newton steps are always eventually taken, and the iterates converge quadratically. For other problems, BFGS steps are eventually used and superlinear convergence is obtained.

Another combination of the Gauss–Newton and quasi-Newton approaches appears in the algorithm implemented in the PORT 3 library. A trust-region approach is used with an approximate Hessian matrix of the form

$$f'(x_k)^T f'(x_k) + S_k,$$

where S_k is a quasi-Newton approximation to the second term in $\nabla^2 r(x)$. Low-rank corrections are applied to S_k at each iteration, together with a scaling strategy that ensures that S_k stays small when the residuals $f_i(x_k)$ are small. At each iteration, a decision is made whether to take the Gauss–Newton step or the step that is computed by including the S_k term.

Other codes that use a quasi-Newton approach that takes advantage of the structure in nonlinear least squares problems in some way are DFNLP, LANCELOT, NLSSOL, and VE10.

NONLINEAR LEAST SQUARES

The algorithms in the VE10 and LANCELOT packages address nonlinear least squares problems where $f'(x)$ is large and sparse. The algorithms make use of the assumption that each f_i is a partially separable function (see Chapter 2 for a definition) to compute compact quasi-Newton approximations to the Hessian matrices $\nabla^2 f_i$. At each iteration a choice is made between a Gauss–Newton step and a step derived from a structured quasi-Newton Hessian approximation. The step is obtained by applying a conjugate gradient method to one of these model choices.

In some applications, it may be necessary to place the bound constraints $l \leq x \leq u$ on the variables x. The resulting problem can be solved with the methods described in Chapter 7, possibly modified to take advantage of the special Hessian approximations that are available for nonlinear least squares problems. Active set methods for handling the bounds form part of the capability of the DFNLP, IMSL, LANCELOT, NAG, NLSSOL, PORT 3, and VE10 codes. An approach based on the gradient-projection method, which is more suitable for large-scale applications, is used by the LANCELOT and VE10 codes.

The PROC NLP codes can be used to solve problems with general linear constraints. The algorithms use active set versions of the Levenberg–Marquardt algorithm, as well as of the hybrid strategy that combines the Gauss–Newton and BFGS quasi-Newton algorithms.

The DFNLP and NLSSOL codes can find minimizers of r subject to general nonlinear constraints. The NLSSOL code uses the same sequential quadratic programming strategy as the general nonlinear programming code NPSOL, but it makes use of the Jacobian matrix $f'(x)$ to compute a starting approximation to the Hessian of the Lagrangian for the constrained problem and to calculate the gradient ∇r. DFNLP also makes use of sequential quadratic programming techniques while exploiting the structure of r in its choice of approximate Hessian.

Notes and References

Nonlinear least squares algorithms are discussed in the books of Bates and Watts [3]; Dennis and Schnabel [13]; Fletcher [15]; Gill, Murray, and Wright [18]; and Seber and Wild [35]. The books by Bates and Watts and by Seber and Wild are written from a statistical point of view. Bates and Watts emphasize applications, while Seber and Wild concentrate on computational methods (this book also has a chapter on software considerations). Björck [7] discusses algorithms for linear least squares problems in a comprehensive survey that covers, in particular, sparse least squares problems and nonlinear least squares.

Chapter 4
Nonlinear Equations

Systems of nonlinear equations arise as constraints in optimization problems, but also arise, for example, when differential and integral equations are discretized. In solving a system of nonlinear equations, we seek a vector $x \in \mathbb{R}^n$ such that

$$f(x) = 0,$$

where f is a mapping from $\mathbb{R}^n$ to $\mathbb{R}^n$. Most algorithms in this section are closely related to algorithms for unconstrained optimization and nonlinear least squares. Indeed, algorithms for systems of nonlinear equations usually proceed by seeking a local minimizer to the problem

$$\min \{\|f(x)\| : x \in \mathbb{R}^n\}$$

for some norm $\|\cdot\|$ in $\mathbb{R}^n$, usually the l_2 norm. This strategy is reasonable, since any solution of $f(x) = 0$ is a global solution of the minimization problem.

Newton's method, modified and enhanced, forms the basis for most of the software used to solve systems of nonlinear equations. Given an iterate x_k, Newton's method computes $f(x_k)$ and the Jacobian matrix $f'(x_k)$, finds a step s_k by solving the system of linear equations

(4.1) $$f'(x_k)s_k = -f(x_k),$$

and then sets $x_{k+1} = x_k + s_k$. Most of the computational cost of Newton's method is associated with two operations: evaluation of $f(x_k)$ and $f'(x_k)$, and the solution of the linear system (4.1). Since

$$f'(x) = \bigl(\partial_1 f(x), \ldots, \partial_n f(x)\bigr),$$

the computation of the ith column of $f'(x)$ requires the partial derivative $\partial_i f(x)$ of f with respect to the ith variable, while the solution of the $n \times n$ linear system (4.1) requires order n^3 operations when $f'(x)$ is dense.

Convergence of Newton's method is guaranteed if x_0 is sufficiently close to the solution x^* and $f'(x^*)$ is nonsingular. Under these conditions the rate of convergence is quadratic; that is,

$$\|x_{k+1} - x^*\| \leq \beta \|x_k - x^*\|^2,$$

for some positive constant β. This rapid local convergence is the main advantage of Newton's method. The disadvantages include the need to calculate the Jacobian matrix and the lack of guaranteed global convergence; that is, convergence from remote starting points.

The software in GAUSS, IMSL, LANCELOT, MATLAB, MINPACK-1, NAG, NITSOL, and OPTIMA attempts to overcome these two disadvantages of Newton's method by allowing approximations to be used in place of the exact Jacobian matrix and by using two basic strategies—trust region and line search—to improve global convergence behavior.

A *trust-region* version of Newton's method takes the view that the linear model $f(x_k) + f'(x_k)s$ of $f(x_k+s)$ is valid only when s is not too large, and thus places a restriction on the size of the step. In a general trust-region method, the Jacobian matrix is replaced by an approximation B_k, and the step s_k is obtained as an approximate solution of the subproblem

$$(4.2) \qquad \min\left\{\|f(x_k) + B_k s\| : \|D_k s\|_2 \leq \Delta_k\right\},$$

where D_k is a scaling matrix and Δ_k is the trust-region radius. The step s_k is accepted if the ratio

$$\rho_k = \frac{\|f(x_k)\| - \|f(x_k + s_k)\|}{\|f(x_k)\| - \|f(x_k) + B_k s_k\|}$$

of the actual-to-predicted decrease in $\|f(x)\|$ is greater than some constant σ_0. (Typically, $\sigma_0 = 10^{-4}$.) If the step is not accepted, the radius Δ_k is decreased and s_k is recomputed. The trust-region radius may also be updated between iterations according to how close the ratio ρ_k is to its ideal value of 1. Trust-region techniques are used in the IMSL, LANCELOT, MINPACK-1, and NAG codes.

Given an iterate x_k and an approximation B_k to the Jacobian matrix $f'(x_k)$, a *line-search* method obtains a search direction d_k by solving the system of linear equations

$$B_k d_k = -f(x_k).$$

The next iterate is then defined as $x_{k+1} = x_k + \alpha_k d_k$, where $\alpha_k > 0$ is chosen by the line-search procedure so that $\|f(x_{k+1})\| < \|f(x_k)\|$. When $B_k = f'(x_k)$, as in Newton's method, d_k is a downhill direction with respect to the l_2 norm, so there is certain to be an $\alpha_k > 0$ such that $\|f(x_{k+1})\|_2 < \|f(x_k)\|_2$. This descent property does not necessarily hold for other choices of B_k, so line-search methods are used only when B_k is either the exact Jacobian or a close approximation to it.

In an ideal line-search Newton method, we would compute the search direction by solving

$$(4.3) \qquad f'(x_k)d_k = -f(x_k)$$

and choose the line-search parameter α_k to minimize the scalar function

$$\phi(\alpha) = \|f(x_k + \alpha d_k)\|_2^2.$$

However, since it is usually too time-consuming to find the α that exactly minimizes ϕ, we usually settle for an approximate solution α_k that satisfies the conditions

$$\phi(\alpha_k) \leq \phi(0) + \mu\alpha_k\phi'(0), \quad |\phi'(\alpha_k)| \leq \eta|\phi'(0)|,$$

where μ and η are two constants with $0 < \mu < \eta < 1$. (Typical values are $\mu = 0.001$ and $\eta = 0.9$.) The first of these conditions ensures that $\|f(x)\|_2^2$ decreases by a significant amount, while the second condition ensures that we move far enough along d_k by insisting on a significant reduction in the size of the gradient. Line-search techniques are implemented in the GAUSS, MATLAB, OPTIMA, and TENSOLVE codes.

For many large-scale problems, determination of the direction d_k consumes most of the computing time. When x_k is still some distance from x^*, there is not much point in obtaining an exact solution to (4.3); an approximate solution often leads to a search direction that is just as useful. This observation is the motivation for the truncated Newton method, a variant of which is implemented in the NITSOL package. In this method, the search direction d_k satisfies

$$\|f(x_k) + f'(x_k)d_k\|_2 \leq \tau_k\|f(x_k)\|_2$$

for some $\tau_k \in (0,1)$. NITSOL finds d_k by using GMRES, an iterative code for nonsymmetric linear equations, and it performs a line search along d_k to obtain the step. The code contains heuristics for adjusting the value of τ_k between iterations. As the solution x^* is approached, τ_k is decreased to zero to ensure superlinear convergence.

When derivatives are not available or are difficult to calculate, the matrix B_k can be either a finite-difference approximation to the Jacobian matrix or a matrix generated by Broyden's method. Approximations based on differences of function values can be obtained by noting that the ith column of $f'(x)$ is approximated by

$$\frac{f(x + h_i e_i) - f(x)}{h_i},$$

where h_i is a small scalar and e_i is the ith unit vector (the ith component of e_i is 1; all other components are zero). The choice of the difference parameter h_i can be a source of difficulty for many codes, particularly if the problem is highly nonlinear or if the function is noisy.

Broyden's method updates the matrix B_k at each iterate so that the new approximation B_{k+1} satisfies the quasi-Newton equation

(4.4) $$B_{k+1}(x_{k+1} - x_k) = f(x_{k+1}) - f(x_k).$$

Given an initial matrix B_0 (often a finite-difference approximation to the Jacobian matrix), Broyden's method generates subsequent matrices by using the updating formula

(4.5) $$B_{k+1} = B_k + \frac{(y_k - B_k s_k)s_k^T}{\|s_k\|_2^2},$$

where
$$s_k = x_{k+1} - x_k, \quad y_k = f(x_{k+1}) - f(x_k).$$

The remarkable feature of Broyden's method is that it is able to generate a reasonable approximation to the Jacobian matrix with no additional evaluations of f. This feature is partially explained by noting that, because of equation (4.4), the updated B_{k+1} mimics the behavior of the true Jacobian along the line joining x_k to x_{k+1}.

The IMSL, MINPACK-1, and NAG codes implement a trust-region method where the matrices B_k are usually calculated by Broyden's method. If the algorithm is not making satisfactory progress, B_k is reset to a finite-difference approximation of the Jacobian and is updated by using (4.5) on subsequent iterations.

TENSOLVE implements a tensor method that goes a step beyond Newton's method by including second-order derivative information from f into its model function. For problems with a dense Jacobian matrix, the storage and cost of the linear algebra operations increase only marginally over Newton's method. The tensor method typically converges more rapidly than Newton's method, particularly when $f'(x)$ is singular at the solution x^*.

Algorithms that use trust-region and line-search strategies require the condition $\|f(x_{k+1})\| \leq \|f(x_k)\|$ to hold for each k. Consequently, they can become trapped in a region in which the function $\|f(\cdot)\|$ has a local minimizer z^* for which $f(z^*) \neq 0$, and they can therefore fail to converge to a solution x^* of $f(x) = 0$. To guarantee convergence to x^* from arbitrary starting points, we need to turn to a more complex class of algorithms known as *homotopy*, or *continuation* methods. These methods are usually slower than line-search and trust-region methods, but they are useful on difficult problems for which a good starting point is hard to find.

Continuation methods define an easy problem for which we know the solution and a path between this easy problem and the hard problem $f(x) = 0$ that we actually wish to solve. The solution to the easy problem is gradually transformed to the solution of the hard problem by tracing this path. The path may be defined by introducing an additional scalar parameter λ into the problem and defining a function

$$h(x, \lambda) = f(x) - (1 - \lambda)f(x_0),$$

where x_0 is a given point in $\mathbb{R}^n$. The problem

(4.6) $$h(x, \lambda) = 0$$

is then solved for values of λ between 0 and 1. When $\lambda = 0$, the solution to (4.6) is clearly $x = x_0$. When $\lambda = 1$, we have that $h(x, 1) = f(x)$, and so the solution of (4.6) coincides with the solution of the original problem $f(x) = 0$.

The path that takes (x, λ) from $(x_0, 0)$ to $(x^*, 1)$ may have turning points at which λ switches from increasing to decreasing and vice versa, so it may not be possible to move from x_0 to x^* by gradually increasing λ from 0 to 1. The algorithms in PITCON and HOMPACK use the more robust approach of expressing both x and λ in terms of a third parameter, s, which represents arc length along the solution path. One of the methods implemented in HOMPACK traces the path by differentiating $h(x, \lambda)$ with respect to s to obtain the ordinary differential equation

$$\partial_x h(x, \lambda)\, x'(s) + \partial_\lambda h(x, \lambda)\, \lambda'(s) = 0.$$

Sophisticated solvers may then be applied to this problem with initial condition $(x(0), \lambda(0)) = (x_0, 0)$ and side condition $\|x'(s)\|_2^2 + \lambda'(s)^2 = 1$. Another method, implemented in both HOMPACK and PITCON, obtains a new iterate (x_{k+1}, λ_{k+1}) from the current iterate (x_k, λ_k) by solving an augmented system of the form

$$\begin{bmatrix} h(x, \lambda) \\ w_k^T x + \mu_k \lambda - t_k \end{bmatrix} = 0.$$

The additional (linear) equation is usually chosen so that (w_k, μ_k) is one of the unit vectors in $\mathbb{R}^{n+1}$, and t_k is a target value for one of the components of (x_{k+1}, λ_{k+1}) whose value has been fixed by a predictor step.

HOMPACK includes additional features such as sparse linear algebra software for large-scale systems and algorithms for finding the solution of systems of polynomial equations. In addition to simple systems of nonlinear equations, PITCON can solve *parametrized* systems, in which λ is an intrinsic parameter rather than an artificial parameter as in the discussion above. The code allows features of the solution path such as bifurcation points and turning points to be calculated accurately.

Notes and References

The book by Dennis and Schnabel [13] gives a good description of algorithms for nonlinear equations. These authors discuss the convergence theory as well as software considerations such as stopping, scaling, and testing.

For information on continuation methods see the survey paper of Watson [36], and the book of Rheinboldt [32]. Watson discusses, in particular, the methods implemented in the HOMPACK software, while Rheinboldt covers the approach used in PITCON. Another source of information on continuation methods is the book of Allgower and Georg [2].

Chapter 5
Linear Programming

Software for linear programming (including network linear programming) consumes more computer cycles than software for all other kinds of optimization problems combined. There is a proliferation of linear programming software with widely varying capabilities and user interfaces. The most recent survey of linear programming software for desktop computers carried out by *OR/MS Today* (19 (1992), pp. 44–59) gave details on 49 packages!

The basic problem of linear programming is to minimize a linear objective function of continuous real variables, subject to linear constraints. For purposes of describing and analyzing algorithms, the problem is often stated in the *standard form*

$$\min \left\{ c^T x : Ax = b, \ x \geq 0 \right\},$$

where $x \in \mathbb{R}^n$ is the vector of unknowns, $c \in \mathbb{R}^n$ is the cost vector, and $A \in \mathbb{R}^{m \times n}$ is the constraint matrix. The feasible region described by the constraints is a polytope, or *simplex*, and at least one member of the solution set lies at a vertex of this polytope.

The *simplex algorithm*, so named because of the geometry of the feasible set, underlies the vast majority of available software packages for linear programming. However, this situation may change in the future, as more software for *interior-point algorithms* becomes available.

The Simplex Method

The simplex method generates a sequence of feasible iterates by repeatedly moving from one vertex of the feasible set to an adjacent vertex with a lower value of the objective function $c^T x$. When it is not possible to find an adjoining vertex with a lower value of $c^T x$, the current vertex must be optimal, and termination occurs.

After its discovery by Dantzig in the 1940s, the simplex method was unrivaled until the late 1980s for its utility in solving practical linear programming problems. Although never observed on practical problems, the poor worst-case behavior of the algorithm—the number of iterations may be exponential in the number of unknowns—led to an ongoing search for algorithms with better computational complexity. This search continued until the late 1970s, when the first polynomial-time algorithm (Khachiyan's ellipsoid method) appeared. Most interior-point methods, which we describe later, also have polynomial complexity.

Algebraically speaking, the simplex method is based on the observation that at least $(n-m)$ of the components of x are zero if x is a vertex of the feasible set. Accordingly, the components of x can be partitioned at each vertex into a set of m basic variables—all nonnegative—and a set of $(n-m)$ nonbasic variables—all zero. If we gather the basic variables into a subvector $x_B \in \mathbb{R}^m$, and the nonbasics into another subvector $x_N \in \mathbb{R}^{n-m}$, we can partition the columns of A as $[B \,|\, N]$, where B contains the m columns that correspond to x_B. (Note that B is a square matrix.)

At each iteration of the simplex method, a basic variable (a component of x_B) is reclassified as nonbasic, and vice versa. In other words, x_B and x_N exchange a component. Geometrically, this swapping process corresponds to a move from one vertex of the feasible set to an adjacent vertex. We therefore need to choose which component of x_N should enter x_B (that is, be allowed to move off its zero bound) and which component of x_B should enter x_N (that is, be driven to zero). In fact, we need make only the first of these choices, since the second choice is implied by the feasibility constraints $Ax = b$ and $x \geq 0$. In selecting the entering component, we note that $c^T x$ can be expressed as a function of x_N alone. We can express x_B in terms of x_N by noting that $Ax = b$ implies that

$$x_B = B^{-1}(b - N x_N).$$

Hence, partitioning c into c_B and c_N in the obvious way, we have

$$c^T x = c_B^T x_B + c_N^T x_N = c_B^T B^{-1} b + \left[c_N - N^T B^{-T} c_B\right]^T x_N.$$

The vector $d_N = c_N - N^T B^{-T} c_B$ is the *reduced-cost vector*. If all components of x_B are strictly positive and some component (say, the ith component) of d_N is negative, we can decrease the value of $c^T x$ by allowing component i of x_N to become positive while adjusting x_B to maintain feasibility. Unless there exist feasible points x that make $c^T x$ arbitrarily negative, the requirement $x_B \geq 0$ imposes an upper bound on $x_{N,i}$, the ith component of x_N.

In principle, we can choose any component $x_{N,i}$ with $d_{N,i} < 0$ as a basic variable. If there are no negative entries in d_N, the current point x is optimal. If there is more than one, we would ideally pick the component that will lead to the largest reduction in $c^T x$ on the current iteration. Heuristics for making this selection are discussed below.

It follows from $Ax = b$ and the fact that the remaining elements of x_N are held at zero that

$$x_B = B^{-1}(b - N_i x_{N,i}),$$

where N_i denotes the column of N that corresponds to $x_{N,i}$. We choose the new value of $x_{N,i}$ to be the largest value that maintains $x_B \geq 0$. To obtain

LINEAR PROGRAMMING

$x_{N,i}$ explicitly, we can rearrange the previous equation to obtain

$$x_{N,i} = \min\left\{\frac{(B^{-1}b)_j}{(B^{-1}N_i)_j} : (B^{-1}N_i)_j > 0\right\}.$$

The index j that achieves the minimum in this formula indicates the basic variable $x_{B,j}$ that is to become nonbasic. If more than one such component achieves the minimum simultaneously, the one with the largest value of $(B^{-1}N_i)_j$ is usually selected.

Most of the computational cost in simplex algorithms arises from the need to compute the vectors $B^{-T}c_B$ and $B^{-1}N_i$ and the need to keep track of the changes in B and B^{-1} resulting from the changes in the basis at each iteration. We could simply recompute and store B^{-1} explicitly after each step. This strategy is undesirable for two reasons. First, the matrix B^{-1} is usually dense even though the original B is sparse; hence, explicit calculation of B^{-1} requires prohibitive amounts of computing time and storage. Second, since B changes only slightly from one iteration to the next, we should be able to update information about B and B^{-1} rather than recompute it anew at every step.

A technique that is used in many commercial codes is to store an LU factorization of B, that is, to keep track of matrices P, Q, L, and U such that

$$B = PLUQ,$$

where P and Q are permutation matrices (identity matrices whose rows have been reordered), while L and U are lower- and upper-triangular matrices, respectively. Since $P^T P = PP^T = I$ and $Q^T Q = QQ^T = I$, we can calculate $z = B^{-T}c_B$ by performing the following sequence of operations:

$$U^T z_1 = Qc_B, \qquad L^T z_2 = z_1, \qquad z = Pz_2.$$

The first two operations are back- and forward-substitutions with triangular matrices, which can be performed efficiently, while the final operation is a simple rearrangement of the elements of z_2. Calculation of $B^{-1}N_i$ proceeds similarly. The permutation matrices P and Q are chosen so that the factorization is reasonably stable and the factors L and U are not too much denser than the original matrix A.

When B is changed by a single column, the factorization can be updated by applying a number of elementary transformations (that is, additions of multiples of one row of the matrix to another row). Rather than applying these transformations explicitly to the existing factors, they usually are stored in compact form. When the storage occupied by the elementary transformations becomes excessive, they are discarded, and the current basis matrix B is refactored from scratch.

We return to strategies for choosing the component i of x_N to enter the basis, an operation that is known as *pricing* in linear programming parlance.

The simplest strategy is to choose i to correspond to the most negative component of the reduced-cost vector d_N. A more refined approach is the *steepest-edge* strategy, in which we choose the edge along which the objective function decreases most rapidly. The extra computation needed to identify the steepest edge is often more than offset by a reduction in the number of iterations, and this strategy is an option in packages such as CPLEX and OSL.

When the linear program is too large for the data to be stored in core memory, the cost of computing the complete reduced-cost vector d_N at each iteration may require too much traffic with secondary storage, and it may take too long. In this situation, a *partial pricing* strategy may be appropriate. This strategy finds only a subvector of d_N and chooses the entering variable from those components that are actually computed. Of course, the subset of indices that defines the subvector of d_N should be changed frequently.

A problem with all of these strategies is that they do not predict the actual decrease in objective function $c^T x$ that will occur on this iteration. It may happen that we are able to move only a short distance along the chosen edge before encountering another vertex, so the reduction may be minimal. A *multiple pricing* strategy selects a small group of columns with negative reduced costs and computes the actual reduction that would be achieved if any one of the corresponding variables entered the basis. This process is expensive, since it requires the calculation of $B^{-1}N_i$ for each candidate i. One of the candidates is chosen, and the remainder are retained as candidates for the next iteration, since the marginal cost of updating the column $B^{-1}N_i$ to correspond to the new basis matrix is not too high. Of course, the candidate list must be refreshed frequently.

The CPLEX, C-WHIZ, FortLP, LAMPS, LINDO, MINOS, OSL, and PC-PROG packages can be used to solve large-scale problems. Each of these packages accepts input in the industry-standard MPS format. Additionally, some have their own customized input format (for example, CPLEX LP format for CPLEX, direct screen input for PC-PROG). Others can be operated in conjunction with modeling languages (CPLEX, LAMPS, MINOS, and OSL interface with GAMS; LINDO and OSL interface with the LINGO modeling language; CPLEX, MINOS, and OSL interface with AMPL). Recently, interfaces between spreadsheet programs and linear programming packages have become available. The What's *Best!* package links a wide range of standard spreadsheets (including Lotus 1-2-3 and Quattro-Pro) to LINDO.

The IMSL and NAG libraries contain simplex-based subroutines. These codes use dense linear algebra techniques to manipulate the basis matrix and hence are suited for small- to medium-scale problems, rather than large-scale problems. The linear programming problem must be specified in the form

$$\min \left\{ c^T x : b_l \leq Ax \leq b_u,\ l \leq x \leq u \right\},$$

where A is an $m \times n$ matrix. The user must set up the problem data and pass it to the appropriate subroutine.

The BQPD package is aimed primarily at quadratic programming problems, but it does solve linear programs as a special case. BQPD can take advantage of sparsity; as with the libraries above, it is the user's responsibility to supply the problem data through subroutine arguments.

The packages LSSOL and QPOPT are aimed at linear least squares or quadratic programming problems but, as part of their capability, they can solve small- to medium-sized linear programs.

Interior-Point Methods

The announcement by Karmarkar [23] in 1984 that he had developed a fast algorithm that generated iterates that lie in the interior of the feasible set (rather than on the boundary, as simplex methods do) opened up exciting new avenues for research in both the computational complexity and mathematical programming communities. Since then, there has been intense research into a variety of methods that maintain strict feasibility of all iterates, at least with respect to the inequality constraints. Although dwarfed in volume by simplex-based packages, interior-point products such as OB1, OSL, and KORBX have emerged and have proven to be competitive with, and often superior to, the best simplex packages, especially on large problems.

We discuss only *primal-dual* interior-point algorithms. Recent research has shown the primal-dual class to be the most promising from a practical point of view, as well as the most amenable to theoretical analysis.

Writing the dual of the standard form linear programming problem as

$$\max \left\{ b^T y : s = c - A^T y \geq 0 \right\},$$

we see that optimality conditions for (x, y, s) to be a primal-dual solution triplet are that

$$Ax = b, \quad A^T y + s = c, \quad SXe = 0, \quad x \geq 0, \quad s \geq 0,$$

where $S = \text{diag}(s_1, s_2, \ldots, s_n)$, $X = \text{diag}(x_1, x_2, \ldots, x_n)$, and e is the vector of all ones. Interior-point algorithms generate iterates (x_k, y_k, s_k) such that $x_k > 0$ and $s_k > 0$. As $k \to \infty$, the equality-constraint violations $\|Ax_k - b\|$ and $\|A^T y_k + s_k - c\|$ and the duality gap $x_k^T s_k$ are driven to zero, yielding a limiting point that solves the primal and dual linear programs.

Primal-dual methods can be thought of as a variant of Newton's method applied to the system of equations formed by the first three optimality conditions. Given the current iterate (x_k, y_k, s_k) and the damping parameter $\sigma_k \in [0, 1]$, the search direction (w_k, z_k, t_k) is generated by solving the linear system

$$Aw = b - Ax_k, \quad A^T z + t = c - A^T y_k - s_k, \quad S_k w + X_k t = -S_k X_k e + \sigma_k \mu_k e,$$

where $\mu_k = x_k^T s_k / n$. The new point is then obtained by setting

$$(x_{k+1}, y_{k+1}, s_{k+1}) \leftarrow (x_k, y_k, s_k) + (\alpha_k^P w_k, \alpha_k^D z_k, \alpha_k^D t_k),$$

where α_k^P and α_k^D are chosen to ensure that $x_{k+1} > 0$ and $s_{k+1} > 0$.

When $\sigma_k = 0$, the search direction is the pure Newton search direction for the nonlinear system $(Ax = b, A^T y + s = c, SXe = 0)$, and the resulting method is an *affine scaling* algorithm. The effect of choosing positive values of σ_k is to orient the step away from the boundary of the nonnegative orthant defined by $x \geq 0$, $s \geq 0$, thus allowing longer step lengths α_k^P, α_k^D to be taken. *Path-following* methods require α_k^P and α_k^D to be chosen so that x_k and s_k are not merely positive but also satisfy the centrality condition $(x_k, s_k) \in C_k$, where

$$C_k = \{(x,s) : x_i s_i \geq \gamma \mu_k,\ i = 1, \ldots, n\}$$

for some $\gamma \in (0, 1)$. The other requirement on α_k^P and α_k^D is that the decrease in μ_k should not outpace the improvement in feasibility (that is, the decrease in $\|Ax_k - b\|$ and $\|A^T y_k + s_k - c\|$). Greater priority is placed on attaining feasibility than on closing the duality gap. It is possible to design path-following algorithms satisfying these requirements for which the sequence $\{\mu_k\}$ converges to zero at a linear rate. Further, the number of iterates required for convergence is a polynomial function of the size of the problem (typically, order n or order $n^{3/2}$). By allowing the damping parameter σ_k to become small as the solution is approached, the method behaves more and more like a pure Newton method, and superlinear convergence can be achieved.

The OB1 software was developed before the convergence analysis outlined above had been derived, so while its performance is excellent on many problems, it does not satisfy the formal requirements of the theoretical analysis, such as membership of the central path neighborhood C_k. In OB1, σ_k is chosen to be uniformly equal to a small constant depending on n, while α_k^P and α_k^D are taken to be 99.5% of the largest steps that can be taken without violating the nonnegativity of x and s.

Most of the computational cost of an interior-point method is associated with the solution of the linear system that defines the search direction. In OB1, elimination is used to obtain a single linear system in z alone, specifically,

$$AX_k S_k^{-1} A^T z = AX_k S_k^{-1} \left[c - A^T y_k - s_k\right] + b - \sigma_k \mu_k AS_k^{-1} e.$$

The coefficient matrix is formed explicitly, except that dense columns of A (which would cause $AX_k S_k^{-1} A^T$ to have many more nonzeros than A itself) are handled separately, and columns that correspond to apparently nonbasic primal variables are eventually dropped. Another useful feature of OB1 is that the user can force the code to switch to the simplex method once the interior-point algorithm has reached the vicinity of a solution.

Recently, interior-point algorithms have been included among the linear programming options in the OSL system from IBM. There is a primal option (based on a primal log barrier function), a primal-dual path-following strategy, and a primal-dual predictor-corrector algorithm. All these solvers have an option for switching over to a simplex solver when the interior-point method indicates that a solution is near at hand.

Notes and References

The description of the simplex method that we have given is based on the standard form of the linear programming problem. This is the standard approach taken in most textbooks; see, for example, Chvátal [8]. The book of Gill, Murray, and Wright [19] describes the simplex method from both the classical perspective and the linearly constrained viewpoint; this book also contains background material on linear algebra. Murtagh [28] gives a more concise, handbook-style treatment. See Chapter 9 for books on network linear programming problems.

No textbook is yet available for interior-point methods, but Gonzaga [21] surveys the theory of interior-point algorithms for linear programming, while Lustig, Marsten, and Shanno [26] describe the code OB1 and its performance. Research on interior-point methods continues at a furious pace, and many of the results that have been proved to date will most likely become obsolete in the near future.

Chapter 6
Quadratic Programming

The quadratic programming problem involves minimization of a quadratic function subject to linear constraints. Most codes use the formulation

(6.1) $$\min \left\{ \tfrac{1}{2} x^T Q x + c^T x : a_i^T x \leq b_i,\ i \in \mathcal{I},\ a_i^T x = b_i,\ i \in \mathcal{E} \right\},$$

where $Q \in \mathbb{R}^{n \times n}$ is symmetric, and the index sets $\mathcal{I}$ and $\mathcal{E}$ specify the inequality and equality constraints, respectively.

The difficulty of solving the quadratic programming problem depends largely on the nature of the matrix Q. In *convex* quadratic programs, which are relatively easy to solve, the matrix Q is positive semidefinite. If Q has negative eigenvalues—*nonconvex* quadratic programming—then the objective function may have more than one local minimizer. An extreme example is the problem

$$\min \left\{ -x^T x\ :\ -1 \leq x_i \leq 1,\ i = 1, \ldots, n \right\},$$

which has a minimizer at any x with $|x_i| = 1$ for $i = 1, \ldots, n$—a total of 2^n local minimizers.

The codes in the BQPD, LINDO, LSSOL, PC-PROG, PORT 3, and QPOPT packages are based on *active set methods*. After finding a feasible point during an initial phase, these methods search for a solution along the edges and faces of the feasible set by solving a sequence of equality-constrained quadratic programming problems. Active set methods differ from the simplex method for linear programming in that neither the iterates nor the solution need be vertices of the feasible set. When the quadratic programming problem is nonconvex, these methods usually find a local minimizer. Finding a *global* minimizer is a more difficult task that is not addressed by the software currently available.

Equality-constrained quadratic programs arise, not only as subproblems in solving the general problem, but also in structural analysis and other areas of application. *Null-space methods* for solving

(6.2) $$\min \left\{ \tfrac{1}{2} x^T Q x + c^T x\ :\ A x = b \right\}$$

find a full-rank matrix $Z \in \mathbb{R}^{n \times m}$ such that Z spans the null space of A. This matrix can be computed with orthogonal factorizations or, in the case of sparse problems, by LU factorization of a submatrix of A, just as in the simplex method for linear programming. Given a feasible vector x_0, we can express any other feasible vector x in the form

(6.3) $$x = x_0 + Zw,$$

for some $w \in \mathbb{R}^m$. Direct computation shows that the equality-constrained subproblem (6.2) is equivalent to the unconstrained subproblem

$$\min \left\{ \tfrac{1}{2} w^T (Z^T Q Z) w + (Q x_0 + c)^T Z w : w \in \mathbb{R}^m \right\}.$$

If the reduced Hessian matrix $Z^T Q Z$ is positive definite, then the unique solution w^* of this subproblem can be obtained by solving the linear system

$$(Z^T Q Z) w = -Z^T (Q x_0 + c).$$

The solution x^* of the equality-constrained subproblem (6.2) is then recovered by using (6.3). Lagrange multipliers can be computed from x^* by noting that the first-order condition for optimality in (6.2) is that there exists a multiplier vector λ^* such that

$$Q x^* + c + A^T \lambda^* = 0.$$

If A has full rank, then

$$\lambda^* = -(A A^T)^{-1} A (Q x^* + c)$$

is the unique set of multipliers. Most codes uses null-space methods. *Range-space methods* for (6.2) can be used when Q is positive definite and easy to invert, for example, diagonal or block-diagonal. In this approach, (x^*, λ^*) is calculated from the formulae

$$\lambda^* = -(A Q^{-1} A^T)^{-1} (b + A Q^{-1} c), \qquad x^* = -Q^{-1}(c + A^T \lambda^*).$$

Although this approach works only for a subclass of problems, there are many applications in which it is useful.

Active set methods for the inequality-constrained problem (6.1) solve a sequence of equality-constrained problems. Given a feasible x_k, these methods find a direction d_k by solving the subproblem

(6.4) $$\min \left\{ q(x_k + d) : a_i^T (x_k + d) = b_i, \ i \in \mathcal{W}_k \right\},$$

where q is the objective function

$$q(x) = \tfrac{1}{2} x^T Q x + c^T x,$$

QUADRATIC PROGRAMMING

and $\mathcal{W}_k$ is a *working set* of constraints. In all cases $\mathcal{W}_k$ is a subset of

$$\mathcal{A}(x_k) = \left\{i \in \mathcal{I} : a_i^T x_k = b_i\right\} \cup \mathcal{E},$$

the set of constraints that are active at x_k. Typically, $\mathcal{W}_k$ either is equal to $\mathcal{A}(x_k)$ or else has one fewer index than $\mathcal{A}(x_k)$.

The working set $\mathcal{W}_k$ is updated at each iteration with the aim of determining the set $\mathcal{A}^*$ of active constraints at a solution x^*. When $\mathcal{W}_k$ is equal to $\mathcal{A}^*$, a local minimizer of the original problem can be obtained as a solution of the equality-constrained subproblem. The updating of $\mathcal{W}_k$ depends on the solution of subproblem (6.4).

Subproblem (6.4) has a solution if the reduced Hessian matrix $Z_k^T Q Z_k$ is positive definite. This is always the case if Q is positive definite. If subproblem (6.4) has a solution d_k, we compute the largest possible step

$$(6.5) \qquad \mu_k = \max\left\{\frac{b_i - a_i^T x_k}{a_i^T d_k} \,:\, a_i^T d_k > 0,\, i \notin \mathcal{W}_k\right\}$$

that does not violate any constraints, and we set $x_{k+1} = x_k + \alpha_k d_k$, where

$$\alpha_k = \min\{1, \mu_k\}.$$

The step $\alpha_k = 1$ would take us to the minimizer of the objective function on the subspace defined by the current working set, but it may be necessary to truncate this step if a new constraint is encountered. The working set is updated by including in $\mathcal{W}_{k+1}$ all constraints active at x_{k+1}.

If the solution to subproblem (6.4) is $d_k = 0$, then x_k is the minimizer of the objective function on the subspace defined by $\mathcal{W}_k$. First-order optimality conditions for (6.4) imply that there are multipliers $\lambda_i^{(k)}$ such that

$$Qx_k + c + \sum_{i \in \mathcal{W}_k} \lambda_i^{(k)} a_i = 0.$$

If $\lambda_i^{(k)} \geq 0$ for $i \in \mathcal{W}_k$, then x_k is a local minimizer of problem (6.1). Otherwise, we obtain $\mathcal{W}_{k+1}$ by deleting one of the indices i for which $\lambda_i^{(k)} < 0$. As in the case of linear programming, various *pricing* schemes for making this choice can be implemented.

If the reduced Hessian matrix $Z_k^T Q Z_k$ is indefinite, then subproblem (6.4) is unbounded below. In this case we need to determine a direction d_k such that $q(x_k + \alpha d_k)$ is unbounded below, using techniques based on factorizations of the reduced Hessian matrix. Given d_k, we compute μ_k as in (6.5) and define $x_{k+1} = x_k + \mu_k d_k$. The new working set $\mathcal{W}_{k+1}$ is obtained by adding to $\mathcal{W}_k$ all constraints active at x_{k+1}.

A key to the efficient implementation of active set methods is the reuse of information from solving the equality-constrained subproblem at the next iteration. The only difference between consecutive subproblems is that the working set grows or shrinks by a single component. Efficient codes perform updates of the matrix factorizations obtained at the previous iteration, rather than calculating them from scratch each time.

The LSSOL package (duplicated in NAG) is specifically designed for convex quadratic programs and linearly constrained linear least squares problems. It is not aimed at large-scale problems; the constraint matrices and the Hessian Q are all specified in dense storage format. The quadratic programming routine in NLPQL has the same properties. IMSL contains codes for dense quadratic programs. If the matrix Q is not positive definite, it is replaced by $Q + \gamma I$, where γ is chosen large enough to force convexity.

BQPD uses a null-space method to solve quadratic programs that are not necessarily convex. The linear algebra operations are performed in a modular way; the user is allowed to choose between sparse and dense matrix algebra. The reduced Hessian matrix is, however, processed as a dense matrix, even when sparse techniques are used to handle Q and the constraints. The code is efficient for large-scale problems when the size of the working set is close to n. LINDO also takes account of sparsity, while MATLAB, PC-PROG, and QPOPT (also available in the NAG library) are designed for dense quadratic programs that are not necessarily convex.

Linear least squares problems are special instances of convex quadratic programs that arise frequently in data-fitting applications. The linear least squares problem

$$\min \left\{ \tfrac{1}{2} \|Cx - d\|_2^2 \ : \ a_i^T x \leq b_i, \ i \in \mathcal{I}, \ a_i^T x = b_i, \ i \in \mathcal{E} \right\},$$

where $C \in \mathbb{R}^{m \times n}$ and $d \in \mathbb{R}^m$, is a special case of problem (6.1); we can see this by replacing Q by $C^T C$ and c by $C^T d$ in (6.1). In general, it is preferable to solve a least squares problem with a code that takes advantage of the special structure of the least squares problem (for example, LSSOL).

Algorithms for solving linear least squares problems tend to rely on null-space active set methods. For a least squares problem the null-space matrix Z can be obtained from the QR decomposition of C; explicit formation of $C^T C$ is avoided, since $C^T C$ is usually less well conditioned than C.

Notes and References

The books of Fletcher [15] and Gill, Murray, and Wright [18] have chapters on the quadratic programming problem. Unconstrained and equality-constrained linear least squares problems have received considerable attention from the linear algebra community. Additional information on these problems can be found in Lawson and Hanson [25], and Golub and Van Loan [20]. Björck's [7] survey covers, in particular, sparse least squares problems and inequality-constrained linear least squares problems.

Chapter 7
Bound-Constrained Optimization

Bound-constrained optimization problems play an important role in the development of software for the general constrained problem because many constrained codes reduce the solution of the general problem to the solution of a sequence of bound-constrained problems. The development of software for this problem, which we state as

$$\min \{f(x) : l \leq x \leq u\},$$

is also important in applications because parameters that describe physical quantities are often constrained to lie in a given range.

Algorithms for the solution of bound-constrained problems seek a local minimizer x^* of f. The standard first-order necessary condition for a local minimizer x^* can be expressed in terms of the *binding* set

$$\mathcal{B}(x^*) = \{i : x_i^* = l_i, \ \partial_i f(x^*) \geq 0\} \cup \{i : x_i^* = u_i, \ \partial_i f(x^*) \leq 0\}$$

at x^* by requiring that

$$\partial_i f(x^*) = 0, \qquad i \notin \mathcal{B}(x^*).$$

There are other ways to express this condition, but this form brings out the importance of the binding constraints. A second-order sufficient condition for x^* to be a local minimizer of the bound-constrained problem is that the first-order condition hold and that $w^T \nabla^2 f(x^*) w > 0$ for all vectors w with

$$w \neq 0, \quad w_i = 0, \ i \in \mathcal{B}_s(x^*),$$

where

$$\mathcal{B}_s(x^*) = \mathcal{B}(x^*) \cap \{i : \partial_i f(x^*) \neq 0\}$$

is the *strictly binding* set at x^*.

Given any set of *free* variables $\mathcal{F}$, we can define the *reduced gradient* and the *reduced Hessian* matrix, respectively, as the gradient of f and the Hessian matrix of f with respect to the free variables. In this terminology, the second-order condition requires that the reduced gradient be zero and that the reduced

Hessian matrix be positive definite when the set $\mathcal{F}$ of free variables consists of all the variables that are not strictly binding at x^*. As we shall see, algorithms for the solution of bound-constrained problems use unconstrained minimization techniques to explore the reduced problem defined by a set $\mathcal{F}_k$ of free variables. Once this exploration is complete, a new set of free variables is chosen with the aim of driving the reduced gradient to zero.

IMSL, LANCELOT, MATLAB, NAG, OPTIMA, PORT 3, TN/TNBC, and VE08 implement quasi-Newton, truncated Newton, and Newton algorithms for bound-constrained optimization. These codes implement line-search and trust-region versions of the unconstrained minimization algorithms of Chapter 2, so the discussion in this chapter is brief, emphasizing the differences between the unconstrained and bound-constrained cases.

A line-search method for bound-constrained problems generates a sequence of iterates by setting $x_{k+1} = x_k + \alpha_k d_k$, where x_k is a feasible approximation to the solution, d_k is a search direction, and $\alpha_k > 0$ is the step. The direction d_k is obtained as an approximate minimizer of the subproblem

$$(7.1) \qquad \min\left\{\nabla f(x_k)^T d + \tfrac{1}{2} d^T B_k d : d_i = 0, \ i \in \mathcal{W}_k\right\},$$

where $\mathcal{W}_k$ is the *working* set and B_k is an approximation to the Hessian matrix of f at x_k. All variables in the working set $\mathcal{W}_k$ are fixed during this iteration, while all other variables are in the free set $\mathcal{F}_k$. We can express this subproblem in terms of the free variables by noting that it is equivalent to the unconstrained problem

$$\min\left\{g_k^T w + \tfrac{1}{2} w^T A_k w : w \in \mathbb{R}^{m_k}\right\},$$

where m_k is the number of free variables, A_k is the matrix obtained from B_k by taking those rows and columns whose indices correspond to the free variables, and g_k is obtained from $\nabla f(x_k)$ by taking the components whose indices correspond to the free variables.

The main requirement on $\mathcal{W}_k$ is that d_k be a feasible direction, that is, $x_k + \alpha d_k$ satisfies the constraints for all $\alpha > 0$ sufficiently small. This is certainly the case if $\mathcal{W}_k = \mathcal{A}(x_k)$, where

$$\mathcal{A}(x) = \{i : x_i = l_i\} \cup \{i : x_i = u_i\}$$

is the set of *active* constraints at x. As long as progress is being made with the current $\mathcal{W}_k$, the next working set $\mathcal{W}_{k+1}$ is obtained by merging $\mathcal{A}(x_{k+1})$ with $\mathcal{W}_k$. This updating process is continued until the function cannot be reduced much further with the current working set. At this point the classical strategy is to drop a constraint in $\mathcal{W}_k$ for which $\partial_i f(x_k)$ has the wrong sign, that is, $i \in \mathcal{W}_k$ but $i \notin \mathcal{B}(x_k)$, where the binding set

$$\mathcal{B}(x) = \{i : x_i = l_i, \ \partial_i f(x) \geq 0\} \cup \{i : x_i = u_i, \ \partial_i f(x) \leq 0\}$$

is defined as before. In general it is advantageous to drop more than one constraint, in the hope that the algorithm will make more rapid progress towards the optimal binding set. However, all dropping strategies are constrained by the requirement that the solution d_k of the subproblem be a feasible direction.

An implementation of a line-search method based on subproblem (7.1) must cater to the situation in which the reduced Hessian matrix A_k is indefinite, because in this case the subproblem does not have a solution. This situation may arise, for example, if B_k is the Hessian matrix or an approximation obtained by differences of the gradient. In this case it is necessary to specify d_k by other means; for example, we can use the modified Cholesky factorization as outlined in Chapter 2.

Quasi-Newton methods for bound-constrained problems update an approximation to the reduced Hessian matrix since, as already noted, only the reduced Hessian matrix is likely to be positive definite. The updating process is not entirely satisfactory because there are situations in which a positive definite update that satisfies the quasi-Newton condition does not exist. Moreover, complications arise because the dimension of the reduced matrix changes when the working set $\mathcal{W}_k$ changes. Quasi-Newton methods are usually beneficial, however, when the working set remains fixed during consecutive iterations.

The choice of line-search parameter α_k is quite similar to the unconstrained case. If subproblem (7.1) has a solution d_k and $x_k + d_k$ violates one of the constraints, then we compute the largest $\mu_k \in (0, 1)$ such that $x_k + \mu_k d_k$ is feasible. A standard strategy for choosing α_k is to seek an $\alpha_k \in (0, \mu_k]$ that satisfies the sufficient decrease and curvature conditions of Chapter 2. We are guaranteed the existence of such an α_k unless μ_k satisfies the sufficient decrease condition and

$$\nabla f(x_k + \mu_k d_k)^T d_k < 0.$$

This situation is likely to happen if, for example, f is strictly decreasing on the line segment $[x_k, x_k + \mu_k d_k]$. In this case it is safe to set $\alpha_k = \mu_k$.

Active set methods have been criticized because the working set changes slowly; at each iteration at most one constraint is added to or dropped from the working set. If there are k_0 constraints active at the initial $\mathcal{W}_0$, but k_s constraints active at the solution, then at least $|k_s - k_0|$ iterations are required for convergence. This property can be a serious disadvantage in large problems if the working set at the starting point is vastly different from the active set at the solution. Consequently, recent investigations have led to algorithms that allow the working set to undergo radical changes at each iteration and to interior-point algorithms that do not explicitly maintain a working set.

The *gradient-projection* algorithm is the prototypical method that allows large changes in the working set at each iteration. Given x_k, this algorithm searches along the piecewise linear path

$$P\left[x_k - \alpha \nabla f(x_k)\right], \qquad \alpha \geq 0,$$

where P is the projection onto the feasible set, and defines

$$x_{k+1} = P\left[x_k - \alpha_k \nabla f(x_k)\right]$$

once it finds a suitable $\alpha_k > 0$. For bound-constrained problems the projection can be easily computed by setting

$$[P(x)]_i = \mathrm{mid}\{x_i, l_i, u_i\},$$

where mid$\{\cdot\}$ is the middle (median) element of a set. The definition and search of α_k have to be done with care because the function

$$\phi(\alpha) = f\Big(P\left[x_k - \alpha \nabla f(x_k)\right]\Big)$$

is only piecewise differentiable.

If properly implemented, the gradient-projection method is guaranteed to identify the active set at a solution in a finite number of iterations. After it has identified the correct active set, the gradient-projection algorithm reduces to the steepest-descent algorithm on the subspace of free variables. As a result, this method is invariably used in conjunction with other methods with faster rates of convergence.

The trust-region algorithms defined in Chapter 2 can be extended to bound-constrained problems. The main difference between the unconstrained and the bound-constrained version is that we now require the step s_k to be an approximate solution of the subproblem

$$\min\Big\{q_k(s) : \|D_k s\| \leq \Delta_k, \; l \leq x_k + s \leq u\Big\},$$

where

$$q_k(s) = \nabla f(x_k)^T s + \tfrac{1}{2} s^T B_k s.$$

An accurate solution to this subproblem is not necessary, at least on early iterations. Instead, we use the gradient-projection algorithm to predict a step s_k^C (the *Cauchy step*) and then require merely that our step s_k satisfy the constraints in the trust-region subproblem with $q_k(s_k) \leq q_k(s_k^C)$. An approach along these lines is used by VE08 and PORT 3. In the bound-constrained code in LANCELOT the trust region is defined by the ℓ_∞-norm and $D_k = I$, yielding the equivalent subproblem

$$\min\Big\{q_k(s) : \max(l - x_k, \Delta_k e) \leq s \leq \min(u - x_k, \Delta_k e)\Big\},$$

where e is the vector of all ones.

The advantage of strategies that combine the gradient-projection method with trust-region methods is that the working set is allowed to change rapidly, and yet eventually settle into the working set for the solution. LANCELOT uses this approach, together with special data structures that exploit the (group) partially separable structure of f, to solve large bound-constrained problems.

Notes and References

The books of Fletcher [15] and Gill, Murray, and Wright [18] contain chapters on the solution of linearly constrained problems, with specific details on the solution of bound-constrained problems. Both Newton and quasi-Newton methods are discussed, but neither book discusses the gradient-projection method, since its use in codes for the solution of large-scale problems is recent. Bertsekas [4, pp. 76–92] has a section on the solution of bound-constrained problems with gradient-projection techniques, while Conn, Gould, and Toint [10] discuss the approach used by the LANCELOT code.

Chapter 8
Constrained Optimization

The general constrained optimization problem is to minimize a nonlinear function subject to nonlinear constraints. Two equivalent formulations of this problem are useful for describing algorithms. They are

(8.1) $$\min \{f(x) : c_i(x) \leq 0, \ i \in \mathcal{I}, \ c_i(x) = 0, \ i \in \mathcal{E}\},$$

where each c_i is a mapping from $\mathbb{R}^n$ to $\mathbb{R}$, and $\mathcal{I}$ and $\mathcal{E}$ are index sets for inequality and equality constraints, respectively; and

(8.2) $$\min\{f(x) : c(x) = 0, \ l \leq x \leq u\},$$

where c maps $\mathbb{R}^n$ to $\mathbb{R}^m$, and the lower- and upper-bound vectors, l and u, may contain some infinite components.

The main techniques that have been proposed for solving constrained optimization problems are reduced-gradient methods, sequential linear and quadratic programming methods, and methods based on augmented Lagrangians and exact penalty functions. Fundamental to the understanding of these algorithms is the Lagrangian function, which for formulation (8.1) is defined as

$$\mathcal{L}(x, \lambda) = f(x) + \sum_{i \in \mathcal{I} \cup \mathcal{E}} \lambda_i c_i(x).$$

The Lagrangian is used to express first-order and second-order conditions for a local minimizer. We simplify matters by stating just first-order necessary and second-order sufficiency conditions without trying to make the weakest possible assumptions.

The first-order necessary conditions for the existence of a local minimizer x^* of the constrained optimization problem (8.1) require the existence of Lagrange multipliers λ_i^* such that

$$\nabla_x \mathcal{L}(x^*, \lambda^*) = \nabla f(x^*) + \sum_{i \in \mathcal{A}^*} \lambda_i^* \nabla c_i(x^*) = 0,$$

where

$$\mathcal{A}^* = \{i \in \mathcal{I} : c_i(x^*) = 0\} \cup \mathcal{E}$$

is the *active set* at x^*, and $\lambda_i^* \geq 0$ if $i \in \mathcal{A}^* \cap \mathcal{I}$. This result requires a *constraint qualification* to ensure that the geometry of the feasible set is adequately captured by a linearization of the constraints about x^*; a standard constraint qualification requires the constraint normals $\nabla c_i(x^*)$ for $i \in \mathcal{A}^*$ to be linearly independent.

The second-order sufficiency condition requires that (x^*, λ^*) satisfy the first-order condition $\nabla_x \mathcal{L}(x^*, \lambda^*) = 0$ and that the Hessian of the Lagrangian

$$\nabla_{xx}^2 \mathcal{L}(x^*, \lambda^*) = \nabla^2 f(x^*) + \sum_{i \in \mathcal{A}^*} \lambda_i^* \nabla^2 c_i(x^*)$$

satisfy $w^T \nabla_{xx}^2 \mathcal{L}(x^*, \lambda^*) w > 0$ for all nonzero w in the set

$$\left\{ w \in \mathbf{R}^n : \nabla c_i(x^*)^T w = 0, \ i \in \mathcal{I}_+^* \cup \mathcal{E}, \ \nabla c_i(x^*)^T w \leq 0, \ i \in \mathcal{I}_0^* \right\},$$

where

$$\mathcal{I}_+^* = \{i \in \mathcal{A}^* \cap \mathcal{I} : \lambda_i^* > 0\}, \qquad \mathcal{I}_0^* = \{i \in \mathcal{A}^* \cap \mathcal{I} : \lambda_i^* = 0\}.$$

This condition guarantees that the optimization problem is well behaved near x^*; in particular, if the second-order sufficiency condition holds, then x^* is a strict local minimizer of the constrained problem (8.1). An important ingredient in the convergence analysis of a constrained algorithm is its behavior in the vicinity of a point (x^*, λ^*) that satisfies the second-order sufficiency condition.

The *sequential quadratic programming* (sequential QP) algorithm is a generalization of Newton's method for unconstrained optimization in that it finds a step away from the current point by minimizing a quadratic model of the problem. A number of packages, including NPSOL, NLPQL, OPSYC, OPTIMA, MATLAB, and SQP, are founded on this approach. In its purest form, the sequential QP algorithm replaces the objective function with the quadratic approximation

$$q_k(d) = \nabla f(x_k)^T d + \tfrac{1}{2} d^T \nabla_{xx}^2 \mathcal{L}(x_k, \lambda_k) d$$

and replaces the constraint functions by linear approximations. For the formulation (8.1), the step d_k is calculated by solving the quadratic subprogram

$$(8.3) \qquad \min\Big\{ q_k(d) : \ c_i(x_k) + \nabla c_i(x_k)^T d \leq 0, \ i \in \mathcal{I} \\ c_i(x_k) + \nabla c_i(x_k)^T d = 0, \ i \in \mathcal{E} \Big\}.$$

The local convergence properties of the sequential QP approach are well understood when (x^*, λ^*) satisfies the second-order sufficiency conditions. If the starting point x_0 is sufficiently close to x^*, and the Lagrange multiplier estimates $\{\lambda_k\}$ remain sufficiently close to λ^*, then the sequence generated by setting $x_{k+1} = x_k + d_k$ converges to x^* at a second-order rate. These assurances cannot be made in other cases. Indeed, codes based on this approach must modify the subproblem (8.3) when the quadratic q_k is unbounded below on the feasible set or when the feasible region is empty.

The Lagrange multiplier estimates that are needed to set up the second-order term in q_k can be obtained by solving an auxiliary problem or by simply using the optimal multipliers for the quadratic subproblem at the previous iteration. Although the first approach can lead to more accurate estimates, most codes use the second approach.

The strategy based on (8.3) makes the decision about which of the inequality constraints appear to be active at the solution internally during the solution of the quadratic program. A somewhat different algorithm is obtained by making this decision prior to formulating the quadratic program. This variant explicitly maintains a working set $\mathcal{W}_k$ of apparently active indices and solves the quadratic programming problem

$$(8.4) \qquad \min\left\{q_k(d) : c_i(x_k) + \nabla c_i(x_k)^T d = 0, \ i \in \mathcal{W}_k\right\}$$

to find the step d_k. The contents of $\mathcal{W}_k$ are updated at each iteration by examining the Lagrange multipliers for the subproblem (8.4) and by examining the values of $c_i(x_{k+1})$ at the new iterate x_{k+1} for $i \notin \mathcal{W}_k$. This approach is usually called the EQP (equality-based QP) variant of sequential QP, to distinguish it from the IQP (inequality-based QP) variant described above.

The sequential QP approach outlined above requires the computation of $\nabla^2_{xx}\mathcal{L}(x_k, \lambda_k)$. Most codes replace this matrix with the BFGS approximation B_k, which is updated at each iteration. An obvious update strategy (consistent with the BFGS update for unconstrained optimization) would be to define

$$s_k = x_{k+1} - x_k, \qquad y_k = \nabla_x \mathcal{L}(x_{k+1}, \lambda_k) - \nabla_x \mathcal{L}(x_k, \lambda_k)$$

and update the matrix B_k by using the BFGS formula

$$B_{k+1} = B_k - \frac{B_k s_k s_k^T B_k}{s_k^T B_k s_k} + \frac{y_k y_k^T}{y_k^T s_k}.$$

However, one of the properties that make Broyden-class methods appealing for unconstrained problems—its maintenance of positive definiteness in B_k—is no longer assured, since $\nabla^2_{xx}\mathcal{L}(x^*, \lambda^*)$ is usually positive definite only in a subspace. This difficulty may be overcome by modifying y_k. Whenever $y_k^T s_k$ is not sufficiently positive, y_k is reset to

$$y_k \leftarrow \theta_k y_k + (1 - \theta_k) B_k s_k,$$

where $\theta_k \in [0, 1)$ is the number closest to 1 such that

$$y_k^T s_k \geq \sigma s_k^T B_k s_k$$

for some $\sigma \in (0, 1)$. The SQP and NLPQL codes use an approach of this type.

The convergence properties of the basic sequential QP algorithm can be improved by using a line search. The choice of distance to move along the direction generated by the subproblem is not as clear as in the unconstrained case, where we simply choose a step length that approximately minimizes f along the search direction. For constrained problems we would like the next iterate not only to decrease f but also to come closer to satisfying the constraints. Often these two aims conflict, so it is necessary to weigh their relative importance and define a *merit* or *penalty function*, which we can use as a criterion for determining whether or not one point is better than another. The ℓ_1 merit function

$$(8.5) \qquad \mathcal{P}_1(x;\nu) = f(x) + \sum_{i \in \mathcal{E}} \nu_i |c_i(x)| + \sum_{i \in \mathcal{I}} \nu_i \max(c_i(x), 0),$$

where $\nu_i > 0$ are penalty parameters, is used in the NLPQL, MATLAB, and SQP codes, while the *augmented Lagrangian* merit function

$$\mathcal{L}_A(x, \lambda; \nu) = f(x) + \sum_{i \in \mathcal{E}} \lambda_i c_i(x) + \tfrac{1}{2} \sum_{i \in \mathcal{E}} \nu_i c_i^2(x) + \tfrac{1}{2} \sum_{i \in \mathcal{I}} \psi_i(x, \lambda; \nu),$$

where

$$\psi_i(x, \lambda; \nu) = \frac{1}{\nu_i} \left\{ \max \left\{ 0, \lambda_i + \nu_i c_i(x) \right\}^2 - \lambda_i^2 \right\},$$

is used in the NLPQL, NPSOL, and OPTIMA codes. The OPSYC code for equality-constrained problems (for which $\mathcal{I} = \emptyset$) uses the merit function

$$f(x) + \sum_{i \in \mathcal{E}} \lambda_i c_i(x) + \left(\sum_{i \in \mathcal{E}} \nu_i c_i^2(x) \right)^{1/2},$$

which combines features of $\mathcal{P}_1$ and $\mathcal{L}_A$.

An important property of the ℓ_1 merit function is that if (x^*, λ^*) satisfies the second-order sufficiency condition, then x^* is a local minimizer of $\mathcal{P}_1$, provided the penalty parameters are chosen so that $\nu_i > |\lambda_i^*|$. Although this is an attractive property, the use of $\mathcal{P}_1$ requires care. The main difficulty is that $\mathcal{P}_1$ is not differentiable at any x with $c_i(x) = 0$. Another difficulty is that although x^* is a local minimizer of $\mathcal{P}_1$, it is still possible for the function to be unbounded below. Thus, minimizing $\mathcal{P}_1$ does not always lead to a solution of the constrained problem.

The merit function $\mathcal{L}_A$ has similar properties. If (x^*, λ^*) satisfies the second-order sufficiency condition and $\lambda = \lambda^*$, then x^* is a local minimizer of $\mathcal{P}_1$, provided the penalty parameters ν_i are sufficiently large. If $\lambda \neq \lambda^*$, then we can say only that $\mathcal{L}_A$ has a minimizer $x(\lambda)$ near x^* and that $x(\lambda)$ approaches x^* as λ converges to λ^*. Note that in contrast to $\mathcal{P}_1$, the merit function $\mathcal{L}_A$ is differentiable. The Hessian matrix of $\mathcal{L}_A$ is discontinuous at any x with $\lambda_i + \nu_i c_i(x) = 0$ for $i \in \mathcal{I}$, but, at least in the case $\mathcal{I}_0^* = \emptyset$, these points tend to occur far from the solution.

CONSTRAINED OPTIMIZATION

The use of these merit functions by NLPQL is typical of other codes. Given an iterate x_k and the search direction d_k, NLPQL sets $x_{k+1} = x_k + \alpha_k d_k$, where the step length α_k approximately minimizes $\mathcal{P}_1(x_k + \alpha d_k; \nu)$. If the merit function $\mathcal{L}_A$ is selected, the step length α_k is chosen to approximately minimize

$$\mathcal{L}_A(x_k + \alpha d_k, \lambda_k + \alpha(\lambda_{k+1} - \lambda_k); \nu),$$

where d_k is a solution of the quadratic programming subproblem (8.3) and λ_{k+1} is the associated Lagrange multiplier.

Rather than simply using these merit functions as a test of suitability for steps generated by the sequential QP algorithm, other algorithms optimize them directly. *Augmented Lagrangian* algorithms are based on successive minimization of the augmented Lagrangian $\mathcal{L}_A$ with respect to x, with updates of λ and possibly ν occurring between iterations. An augmented Lagrangian algorithm for the constrained optimization problem (8.2) computes x_{k+1} as an approximate minimizer of the subproblem

$$\min \{\mathcal{L}_A(x, \lambda_k; \nu_k) : l \leq x \leq u\},$$

where

$$\mathcal{L}_A(x, \lambda; \nu) = f(x) + \sum_{i \in \mathcal{E}} \lambda_i c_i(x) + \tfrac{1}{2} \sum_{i \in \mathcal{E}} \nu_i c_i^2(x)$$

includes only the equality constraints. Updating of the multipliers usually takes the form

$$\lambda_i \leftarrow \lambda_i + \nu_i c_i(x_k).$$

This approach is relatively easy to implement because the main computational operation at each iteration is minimization of the smooth function $\mathcal{L}_A$ with respect to x, subject only to bound constraints. A large-scale implementation of the augmented Lagrangian approach can be found in the LANCELOT package, which solves the bound-constrained subproblem by using special data structures to exploit the (group partially separable) structure of the underlying problem. The OPTIMA and OPTPACK libraries also contain augmented Lagrangian codes.

Reduced-gradient algorithms avoid the use of penalty parameters by searching along curves that stay near the feasible set. Essentially, these methods take the formulation (8.2) and use the equality constraints to eliminate a subset of the variables, thereby reducing the original problem to a bound-constrained problem in the space of the remaining variables. If x_B is the vector of eliminated or *basic* variables, and x_N is the vector of *nonbasic* variables, then $x_B = h(x_N)$, where the mapping h is defined implicitly by the equation

$$c[h(x_N), x_N] = 0.$$

(We have assumed that the components of x have been arranged so that the basic variables come first.) In practice, $x_B = h(x_N)$ can be recalculated using

Newton's method whenever x_N changes. Each Newton iteration has the form

$$x_B \leftarrow x_B - \partial_B c(x_B, x_N)^{-1} c(x_B, x_N),$$

where $\partial_B c$ is the Jacobian matrix of c with respect to the basic variables. The original constrained problem is now transformed into the bound-constrained problem

$$\min\{f(h(x_N), x_N) : l_N \leq x_N \leq u_N\}.$$

Algorithms for this reduced subproblem subdivide the nonbasic variables into two categories. These are the *fixed* variables x_F, which usually include most of the variables that are at either their upper or lower bounds and that are to be held constant on the current iteration, and the *superbasic* variables x_S, which are free to move on this iteration. The standard reduced-gradient algorithm, implemented in CONOPT, searches along the steepest-descent direction in the superbasic variables. The generalized reduced-gradient codes GRG2 and LSGRG2 use more sophisticated approaches. They either maintain a dense BFGS approximation of the Hessian of f with respect to x_S or use limited-memory conjugate gradient techniques. MINOS also uses a dense approximation to the superbasic Hessian matrix. The main difference between MINOS and the other three codes is that MINOS does not apply the reduced-gradient algorithm directly to problem (8.1), but rather uses it to solve a linearly constrained subproblem to find the next step. The overall technique is known as a *projected augmented Lagrangian* algorithm.

Operations involving the inverse of $\partial_B c(x_B, x_N)$ are frequently required in reduced-gradient algorithms. These operations are facilitated by an *LU* factorization of the matrix. GRG2 performs a dense factorization, while CONOPT, MINOS, and LSGRG2 use sparse factorization techniques, making them more suitable for large-scale problems.

When some of the components of the constraint functions are linear, most algorithms aim to retain feasibility of all iterates with respect to these constraints. The optimization problem becomes easier in the sense that there is no curvature term corresponding to these constraints that must be accounted for and, because of feasibility, these constraints make no contribution to the merit function. Numerous codes, such as NPSOL, MINOS and some routines from the NAG library, are able to take advantage of linearity in the constraint set. Other codes, such as those in the IMSL, PORT 3, and PROC NLP libraries, are specifically designed for linearly constrained problems. The IMSL codes are based on a sequential quadratic programming algorithm that combines features of the EQP and IQP variants. At each iteration, this algorithm determines a set $\mathcal{N}_k$ of near-active indices defined by

$$\mathcal{N}_k = \{i \in \mathcal{I} : c_i(x_k) \geq -\tau_i\},$$

where the tolerances τ_i tend to decrease on later iterations. The step d_k is obtained by solving the subproblem

$$\min\{q_k(d) : c_i(x_k + d_k) = 0, \ i \in \mathcal{E}, \ c_i(x_k + d_k) \leq c_i(x_k), \ i \in \mathcal{N}_k\},$$

where
$$q_k(d) = \nabla f(x_k)^T d + \tfrac{1}{2} d^T B_k d,$$

and B_k is a BFGS approximation to $\nabla^2 f(x_k)$. This algorithm is designed to avoid the short steps that EQP methods sometimes produce, without taking many unnecessary constraints into account, as IQP methods do.

Finally, we mention *feasible sequential quadratic programming* algorithms, which, as their name suggests, constrain all iterates to be feasible. They are more expensive than standard sequential QP algorithms, but they are useful when the objective function f is difficult or impossible to calculate outside the feasible set, or when termination of the algorithm at an infeasible point (which may happen with most algorithms) is undesirable. The code FSQP solves problems of the form

$$\min \{f(x) : c(x) \le 0, \ Ax = b\}.$$

In this algorithm, the step is defined as a combination of the sequential QP direction, a strictly feasible direction (which points into the interior of the feasible set) and, possibly, a second-order correction direction. This mix of directions is adjusted to ensure feasibility while retaining fast local convergence properties. Feasible algorithms have the additional advantage that the objective function f can be used as a merit function, since, by definition, the constraints are always satisfied. FSQP also solves problems in which f is not itself smooth, but is rather the maximum of a finite set of smooth functions $f_i : \mathbb{R}^n \to \mathbb{R}$.

Notes and References

The books of Fletcher [15] and Gill, Murray, and Wright [18] are good general references. Fletcher's book contains discussions of constrained optimization theory and sequential quadratic programming. Gill, Murray, and Wright discuss reduced-gradient methods, including the algorithm implemented in MINOS. Bertsekas [4] has an excellent treatment of augmented Lagrangian algorithms. For recent surveys of nonlinear programming algorithms, see the papers of Coleman [9]; Conn, Gould, and Toint [11]; Gill, Murray, Saunders, and Wright [17]; and Hager, Horst, and Pardalos [22].

Chapter 9
Network Optimization

Network optimization problems are reputed to be the most prevalent of all optimization problems in practice. They arise naturally in many applications that involve distribution, transportation, or communication, since the relationships between the components of these applications are often defined by a graph consisting of *nodes* and *arcs*, where each arc joins two nodes.

Linear network problems are a special case of linear programming problems in which each algorithm is most easily explained with reference to the node/arc graph that defines the problem, rather than by the language of linear algebra that was used in Chapter 5. For instance, as we see below, the simplex operations of pricing and pivoting are usually interpreted in terms of adjustment of *flows* on the arcs and *prices* at each node, instead of the language of Lagrange multipliers and *LU* factorization that was used to discuss general linear programming.

In the formulation of network problems, the nodes in the node set $\mathcal{N}$ are numbered by $i = 1, 2, \ldots, n$, and the elements of the arc set $\mathcal{A}$ are ordered pairs (i, j), where i is the origin node and j the destination node of the arc. The unknowns x_{ij} are the flow on arc (i, j). For example, if c_{ij} is the unit cost for flow x_{ij}, then the *minimum-cost network flow* problem has the form

$$\min \sum_{(i,j) \in \mathcal{A}} c_{ij} x_{ij},$$

subject to the constraints

$$\sum_{(i,j) \in \mathcal{A}} x_{ij} - \sum_{(j,i) \in \mathcal{A}} x_{ji} = s_i, \quad 1 \leq i \leq n, \quad l_{ij} \leq x_{ij} \leq u_{ij}, \quad (i,j) \in \mathcal{A}.$$

The equality constraints arise because we specify the net outflow s_i for each node. Note that if these so-called flow-conservation constraints are written in the concise form $Ax = s$, then the coefficient matrix A has exactly two nonzeros per column: the column corresponding to arc (i, j) contains a $+1$ in row i and a -1 in row j. The bounds on the flows are referred to as capacity constraints.

The *maximum-flow* and *assignment* problems are important special cases of the minimum-cost network flow problem. In the maximum-flow problem we aim to maximize the total flow between a source node and a sink node subject to the capacity constraints. In the assignment problem, the network is a bipartite graph involving two sets $\mathcal{N}_1$ and $\mathcal{N}_2$ of nodes, and each arc connects a node in $\mathcal{N}_1$ to a node in $\mathcal{N}_2$. The problem is to find a one-to-one matching between the persons $\mathcal{N}_1$ and the objects $\mathcal{N}_2$ such that the sum of costs c_{ij} for the associated arcs is minimized.

A specialized version of the simplex algorithm of Chapter 5 has proved to be an efficient and popular method for solving network optimization problems. Each basis matrix (denoted by B in Chapter 5) corresponds to a *spanning tree* of the graph, that is, a connected subgraph consisting of all n nodes but just $n-1$ of the arcs. It is not necessary to factor the basis matrix. Not only is B sparse, but a leading $(n-1) \times (n-1)$ submatrix is triangular, so that solutions to systems of the form $Bz = y$ and $B^T z = y$ can be obtained cheaply by back- and forward-substitution.

The Lagrange multipliers are commonly referred to as *prices* and denoted by p_i, $i = 1, \ldots, n$. Optimality conditions dictate that if the arc (i, j) appears in the spanning tree for the optimal basis, the price difference $p_i - p_j$ between the nodes equals the unit cost c_{ij} for flow along that arc.

The fundamental operation of the simplex method—pivoting, or the replacement of a basic variable by a nonbasic variable—can also be explained in the language of the problem graph. The new nonbasic variable to be introduced is chosen by comparing the values of $p_i - p_j$, x_{ij}, l_{ij}, and u_{ij} for a subset of the nonbasic arcs. As in the case of general linear programming, a complete search of all nonbasic arcs is considered to be too expensive for large problems. The chosen arc (i, j) is added to the spanning tree, which already contains a path between nodes i and j (that is, an ordered sequence $(i, k_1), (k_1, k_2), \ldots, (k_m, j)$ of arcs in $\mathcal{A}$). Some of the flow that previously traveled by this path is rerouted through the newly introduced arc to preserve the flow-conservation constraints. The amount of flow that can be so transferred is limited by the capacity constraints on the newly introduced arc and the arcs along the path.

When the new arc is chosen properly, pivoting usually results in a decrease in the objective function. It is also possible for the objective function to remain unaltered; this situation can lead to *cycling*, where the algorithm repeatedly returns to the same active set. Simple devices to avoid cycling are present in most practical codes.

The simplex method for minimum-cost network flow problems is implemented in an algorithm of Kennington and Helgason, available in the NETFLOW collection. The CPLEX system, available both as a self-contained system and as a C routine that can be called from the user's programs, can handle network problems with unlimited additional constraints on the flows by using a simplex approach. IBM's OSL system also implements a simplex algorithm.

A somewhat different algorithm of recent origin is the *relaxation algorithm*. To describe this method, we need to formulate the dual of the minimum-cost

NETWORK OPTIMIZATION

problem as

$$\max\left\{\sum_{(i,j)\in\mathcal{A}} q_{ij}(p_i - p_j) + \sum_{i=1}^{n} s_i p_i\right\},$$

where the functions $q_{ij}: \mathbb{R} \to \mathbb{R}$ are defined by

$$q_{ij}(p_i - p_j) = \min_{l_{ij} \leq x_{ij} \leq u_{ij}} (c_{ij} + p_j - p_i)x_{ij}.$$

Although there are no constraints on p, the dual problem is not trivial to solve because the functions $q_{ij}(\cdot)$ are *nonsmooth*; a plot of q_{ij} contains ridges and kinks.

The relaxation algorithm starts with a primal-dual pair (x, p) that satisfies the capacity constraints and the *complementary slackness* conditions

$$\begin{aligned} x_{ij} &= l_{ij} \quad \text{if} \quad p_i - p_j < c_{ij} \\ x_{ij} &\in [l_{ij}, u_{ij}] \quad \text{if} \quad p_i - p_j = c_{ij} \\ x_{ij} &= u_{ij} \quad \text{if} \quad p_i - p_j > c_{ij}, \end{aligned}$$

but not necessarily the flow-conservation constraints. The method aims simultaneously to resolve the discrepancy in these constraints (by adjusting x) and to work towards a maximum of q (by adjusting p). Price adjustments are performed by identifying a subset $\mathcal{S}$ of the nodes such that the directional derivative of q in the direction $d_{\mathcal{S}}$ defined by

$$(d_{\mathcal{S}})_i = \begin{cases} 1 & \text{if } i \in \mathcal{S} \\ 0 & \text{if } i \in \mathcal{S} \end{cases}$$

is positive. If such a direction is found, the algorithm moves along it, stopping when the maximum of q along this direction is identified. The other possible outcome of each iteration is that an *augmenting path* is found, along which the flow of all arcs can be changed to reduce the infeasibility in the flow-conservation equation at each node along the path. The LNOS package contains an implementation of the relaxation algorithm.

The *auction algorithm* is most easily motivated with reference to the assignment problem. As specified earlier, the assignment problem is to match each person $i \in \mathcal{N}_1$ with an object $j_i \in \mathcal{N}_2$ such that (i, j_i) belongs to the arc set $\mathcal{A}$ and the sum

$$\sum_{i=1}^{n} c_{i,j_i}$$

is minimized. A price p_j is associated with each object $j \in \mathcal{N}_2$, and a matching (i, j_i), $i = 1, \ldots, n$ is optimal if

$$j_i = \arg\max_{j \in A(i)} (c_{ij} - p_j),$$

where $A(i)$ is the set of objects j such that $(i, j) \in \mathcal{A}$. In the auction algorithm, some of the persons i that have not yet been tentatively assigned an object bid for the objects j_i that attain the maximum value of $c_{ij} - p_j$, for the *current* set of prices p_j. Person i bids by offering to raise the price of object j_i to make it slightly less desirable than the second-best object, that is, the object $j \in \mathcal{N}_2$ that achieves the second-largest value of $c_{ij} - p_j$. (Note that an increase in p_{j_i} results in a decrease in $c_{i,j_i} - p_{j_i}$, so j_i will no longer attain the maximum when p_{j_i} increases too much.) At the end of this bidding phase, some objects may have received more than one bid. The highest is accepted, and a tentative pairing is arranged between successful bidders and objects, possibly breaking some previously arranged pairings. The algorithm terminates once each object has received a bid from at least one person.

Because person i bids slightly more than is needed to make them indifferent between the objects that achieve the two largest values of $c_{ij} - p_j$, the final pairings produced by this algorithm are nearly, but not exactly, optimal.

Versions of the auction algorithm in which more than one unassigned person is allowed to bid simultaneously obviously lend themselves to parallel implementation. There is also a *reverse-auction* algorithm, in which objects bid for persons on the basis of profits π_i (rather than prices) associated with each person. Extensions of the auction algorithm for the shortest-path and minimum-cost flow problem have also been developed.

The LNOS package contains implementations of the forward- and forward/reverse-auction algorithm for assignment problems, and also an auction code for shortest-path problems.

Codes are also available for generalizations of the linear minimum-cost flow problem described above. In some applications, a nonlinear objective function of the form

$$\sum_{(i,j)\in \mathcal{A}} c_{ij}(x_{ij})$$

may take the place of the linear function. The code LSNNO combines elements of gradient-projection techniques and truncated Newton/quasi-Newton algorithms (see Chapters 2 and 7) to solve problems with nonlinear cost functions. GENOS allows both a nonlinear cost function and generalized network constraints to be present, obtaining the solution by a truncated Newton technique.

Notes and References

The book by Kennington and Helgason [24] contains both an excellent introduction to network optimization and a cogent description of the network simplex method. In particular, Appendix F of their book describes their code from the NETFLO collection.

The recent book of Bertsekas [5] contains a more advanced survey of the area. Much of this chapter is based on the material in this book. Simplex methods, dual-ascent methods (including the relaxation method), and auction algorithms are all described. This book also contains listings of the major algorithms (see the description of LNOS). More detail on auction algorithms appears in Bertsekas [6].

The recent book of Ahuya, Magnanti and Orlin [1] also covers linear network optimization algorithms. An interesting feature of this book is the extensive discussion of network optimization applications. For information on nonlinear network models, algorithms, and applications, see the survey papers of Dembo, Mulvey, and Zenios [12]; and Florian [16].

Chapter 10
Integer Programming

In many applications, the solution of an optimization problem makes sense only if certain of the unknowns are integers. *Integer linear programming* problems have the general form

(10.1) $$\min\left\{c^T x : Ax = b,\ x \geq 0,\ x \in Z^n\right\},$$

where Z^n is the set of n-dimensional integer vectors. In *mixed-integer linear programs*, some components of x are allowed to be real. We restrict ourselves to the pure integer case, bearing in mind that the software can also handle mixed problems with little additional complication of the underlying algorithm.

Integer programming problems, such as the *fixed-charge network flow* problem and the famous *traveling salesman* problem, are often expressed in terms of binary variables. The fixed-charge network problem modifies the minimum-cost network flow paradigm of Chapter 9 by adding a term $f_{ij}y_{ij}$ to the cost, where the binary variable y_{ij} is set to 1 if arc (i,j) carries a nonzero flow x_{ij}; it is set to zero otherwise. In other words, there is a fixed overhead cost for using the arc at all. In the traveling salesman problem we need to find a tour of a number of cities that are connected by directed arcs, so that each city is visited once and the time required to complete the tour is minimized. One binary variable is assigned to each directed arc; a variable x_{ij} is set to 1 if city i immediately follows city j on the tour, and to zero otherwise.

Although a number of algorithms have been proposed for the integer linear programming problem, the *branch-and-bound* technique is used in almost all of the software in our survey. This technique has proven to be reasonably efficient on practical problems, and it has the added advantage that it solves continuous linear programs as subproblems, that is, linear programming problems without integer restrictions.

The branch-and-bound technique can be outlined in simple terms. An *enumeration tree* of continuous linear programs is formed, in which each problem has the same constraints and objective as (10.1) except for some additional bounds on certain components of x. At the root of this tree is the problem (10.1) with the requirement $x \in Z^n$ removed. The solution x to this root problem will not, in general, have all integer components. We now choose some noninteger solution component x_j and define I_j to be the integer part of x_j, that is, $I_j = \lfloor x_j \rfloor$. This gives rise to two subproblems. The left-child problem has the additional constraint $x_j \leq I_j$, whereas in the right-child problem we impose $x_j \geq I_j + 1$. This *branching* process can be carried out recursively; each of the two new problems will give rise to two more problems when we branch on one of the noninteger components of their solution. It follows from this construction that the enumeration tree is binary.

Eventually, after enough new bounds are placed on the variables, integer solutions $x \in Z^n$ are obtained. The value Z_{opt} of $c^T x$ for the best integer solution found so far is retained and used as a basis for pruning the tree. If the continuous problem at one of the nodes of the tree has a final objective value greater than Z_{opt}, so do all of its descendants, since they have smaller feasible regions and hence even larger optimal objective values. The branch emanating from such a node cannot give rise to a better integer solution than the one obtained so far, so we consider it no further; that is, we prune it. Pruning also occurs when we have added so many new bounds to some continuous problem that its feasible region disappears altogether.

The preferred strategy for solving the node problems in the enumeration tree is of the depth-first type: When two child nodes are generated from the current node, we choose one of these two children as the next problem to solve. One reason for using this strategy is that on practical problems the optimal solution usually occurs deep in the tree. There is also a computational advantage: If the dual simplex algorithm is used to solve the linear program at each node, the solution of the child problem can be found by a simple update to the basis matrix factorization obtained at the parent node. The linear algebra costs are trivial.

Two important questions remain: How do we select the noninteger component j on which to branch, and how do we choose the next problem to solve if the branch we are currently working on is pruned? The answer to both questions depends on maintaining a lower bound $\underline{Z}^I$ on the objective value for the continuous linear program at node I, and an estimate Z^I_{int} of the objective value of the best integer solution for the problem at node I. Both values can be calculated when the problem at node I is generated as the child of another problem. After the current branch has reached a dead end, two common strategies for selecting the next problem to solve are to choose the one for which $\underline{Z}^I$ is least or to choose the one for which Z^I_{int} is least. Other strategies use some criterion function that combines $\underline{Z}^I$, Z^I_{int}, and Z_{opt}.

In many codes, the user is allowed to specify a *branching order* to guide the choice of components on which to branch. By the nature of the problem, some components of x may be more significant than others; the algorithms can use this knowledge to make branching decisions that lead to the solution more quickly. In the absence of such an ordering, the *degradations* in the objective value are estimated by forcing each component in turn to its next highest or next lowest integer. The branching variable is often chosen to be the one for which the degradation is greatest.

The CPLEX, FortLP, LAMPS, LINDO, MIP III, OSL, and PC-PROG packages use the branch-and-bound technique to solve mixed-integer linear programs. The NAG Fortran Library (Chapter H) contains a branch-and-bound subroutine, and also an earlier implementation of a cutting plane algorithm due to Gomory. (The latter code is scheduled for removal from the library in the near future.) GAMS interfaces with a number of mixed-integer linear programming solvers, and even with a mixed-integer nonlinear programming solver. LINDO has two other front-end systems: LINGO provides a modeling interface to it, while What's*Best!* provides a variety of spreadsheet interfaces.

The CPLEX integer programming system can be used either as a stand-alone system or as a subroutine that is called from the user's code. CPLEX also implements a *branch-and-cut* strategy, in which the bounds on optimal objective values are tightened by adding additional inequality constraints $Fx \leq f$ to the problem. The matrix F and vector f are chosen so that all integer vectors x satisfying the original constraints $Ax = b$, $x \geq 0$, also satisfy $Fx \leq f$. These extra constraints (*cuts*) have the effect of reducing the size of the set of real vectors that is being considered at each node.

The Q01SUBS package contains routines to solve the quadratic zero-one programming problem

$$\min \left\{ \tfrac{1}{2} x^T Q x + c^T x : x_i \in \{0,1\}, \ i = 1, \ldots, n \right\},$$

where Q may be indefinite, while QAPP solves the assignment problem with a quadratic objective. Both algorithms use a branch-and-bound methodology similar to the techniques described above.

Notes and References

The book by Nemhauser and Wolsey [29] provides a comprehensive guide to integer programming, while their review paper, [30], is a more concise overview. For a survey of linear and integer programming, with emphasis on complexity analysis, see Schrijver's book [34].

Chapter 11
Miscellaneous Optimization Problems

The optimization problems and software discussed in the preceding chapters are mainly for objective functions that are continuously differentiable. This assumption of continuous differentiability is violated in the optimization areas discussed below.

Minimization of structured nonsmooth functions

The DFNLP and FSQP packages solve optimization problems with an objective function f_0 of the form
$$f_0(x) = \|f(x)\|$$
for some smooth function $f : \mathbb{R}^n \to \mathbb{R}^m$, where $\|\cdot\|$ is either the ℓ_1 or the ℓ_∞ norm. These problems can be transformed into constrained optimization problems by introducing additional variables; this approach is used in the DFNLP package.

Minimization of general nonsmooth functions

The BT package can be used to minimize a nonsmooth (locally Lipschitz) function f subject to linear constraints. The code uses a bundle method, in which the generalized gradient at each iterate is approximated by information collected at previous iterates.

Multiobjective minimization

The problem is to optimize simultaneously a collection of objective functions (in some sense) subject to certain constraints. User interaction with the optimization system is usually required to determine the most appropriate combination of the various objectives into a single objective. Once this is done, the resulting problem can be solved with standard techniques. MATLAB and the packages VIG and VIMDA contain routines for problems in which the constraints and objective functions are linear.

Part II
SOFTWARE PACKAGES

Software Classification

Part II consists of the descriptions of software gathered during our survey of numerical optimization software. The descriptions are ordered alphabetically and accompanied by lists of pertinent references; for convenience, we also provide below a list in which they are grouped according to the problem classification of Part I. Because of the overlapping nature of our problem classification, most of the packages can be used to solve more than one class of problem. For instance, a nonlinear programming code could also be used to solve a bound-constrained problem, but it would typically sacrifice some efficiency by handling the bound constraints as general nonlinear constraints. In each of the classes below, we list only those packages that contain special features for handling problems in that class. For example, listed bound-constrained optimization codes either make use of algorithmic techniques described in Chapter 7, or else take advantage of the special nature of the bound constraints in performing linear algebra computations.

Besides listing software collections that are specifically aimed at optimization problems, we also mention two numerical software libraries—IMSL and NAG. We have also listed information on a number of systems that provide a convenient user interface to optimization algorithms. In all cases, additional information can be found under the individual listings for each package, library, or system.

Unconstrained Optimization

BTN, GAUSS, LANCELOT, LBFGS, M1QN2/M1QN3, OPTIMA, PORT 3, PROC NLP, TENMIN, TN, TNPACK, UNCMIN, VE08.

Nonlinear Least Squares

DFNLP, LANCELOT, MINPACK-1, NLSFIT, NLSSOL, ODRPACK, PORT 3, PROC NLP, TENSOLVE, VE10.

Nonlinear Equations

GAUSS, HOMPACK, LANCELOT, MINPACK-1, NITSOL, OPTIMA, PITCON.

Linear Programming

CPLEX, C-WHIZ, FortLP, GAUSS, LAMPS, LINDO, LPsolver, MINOS, OB1, OSL, PC-PROG, PORT 3, QPOPT, What's*Best!*

Quadratic Programming

BQPD, LINDO, LSSOL, OSL, PC-PROG, PORT 3, QPOPT.

Bound-Constrained Optimization

LANCELOT, MODULOPT, OPTIMA, PORT 3, PROC NLP, TN/TNBC, VE08.

Constrained Optimization

CONOPT, DOT, FSQP, GINO, GRG2, LANCELOT, LSGRG2, MINOS, NLPQL, NLPQLB, NLPSPR, NPSOL, OPSYC, OPTIMA, OPTPACK, SQP.

Network Optimization

GENOS, LNOS, LSNNO, NETFLO, NETSOLVE, OSL.

Integer Programming

CPLEX, FortLP, LAMPS, LINDO, MIPIII, OSL, PC-PROG, Q01SUBS, QAPP.

Miscellaneous Problems

Nondifferentiable Optimization—BT.

Multiple Criteria Optimization—VIG, VIMDA.

Engineering Design—CONSOL-OPTCAD, DOC, GENESIS.

PC Software Collections—CNM, NLPE, OptiA.

Libraries with Optimization Capabilities

The IMSL (Fortran and C versions) and NAG (Fortran and C versions) libraries have extensive optimization capabilities. There is often an easy-to-use version of the subroutine with a short calling sequence, and a version of the subroutine with an extensive calling sequence designed for maximum flexibility.

Optimization Systems/Modeling Languages

An optimization system has a programming language for formulating the optimization problem, as well as capabilities for solving the problem. In addition, the system should be able to interface with a standard Fortran or C program. Optimization systems in our collection include AMPL, GAMS, LINGO, MATLAB, SAS, and SPEAKEASY.

SOFTWARE PACKAGES

AMPL

Areas covered by the software

AMPL is a modeling language for mathematical programming problems. It offers an interactive command environment for setting up and solving these problems. A flexible interface enables several solvers to be available at once and allows a user to switch between solvers and to select options that may improve solver performance. Once optimal solutions have been found, they are expressed in terms of the modeler's notation so that they can be viewed and analyzed. All of the general set and arithmetic expressions of AMPL can also be used for displaying data and results; a variety of options are available to format data for browsing on a screen, printing reports, or preparing input to other programs.

Hardware/software environment

The book referenced below is packaged with a 300×300 MS-DOS version of the AMPL language processor and the MINOS and CPLEX optimizers. This bundle encompasses linear, network, integer, and nonlinear programming and is suitable for instructional use or evaluation of the software. A variety of other combinations of operating systems and optimizers are also supported.

Contact address

Software is distributed in conjunction with the book referenced below, which contains a complete tutorial and reference for AMPL. The publisher is

> The Scientific Press
> 651 Gateway Blvd., Suite 1100
> South San Francisco, CA 94080-7014
> Phone: (800) 451-5409 (outside California)
> Phone: (415) 583-8840 (inside California)
> Fax: (415) 583-6371

Reference

R. Fourer, D. M. Gay, and B. W. Kernighan, *AMPL: A Modeling Language for Mathematical Programming*, The Scientific Press, San Francisco, CA, 1993.

BQPD

Areas covered by the software

BQPD solves quadratic programming problems. A general form of the problem is solved that allows upper and lower bounds on all variables and constraints. If the Hessian matrix Q is positive definite, then a global solution is found. The method can also be used when Q is indefinite, in which case a Kuhn–Tucker point that is usually a local solution is found.

Basic algorithms

The code implements a null-space active set method with a technique for resolving degeneracy that guarantees that cycling does not occur even when roundoff errors are present. Feasibility is obtained by minimizing a sum of constraint violations. The Devex method for avoiding near-zero pivots is used to promote stability. The matrix algebra is implemented in such a way that advantage can be taken of sparse factors of the basis matrix. Factors of the reduced Hessian matrix are stored in a dense format, an approach that is most effective when the number of free variables is relatively small. The user is asked to supply a subroutine to evaluate the Hessian matrix Q, so that sparsity in Q can be exploited. An extreme case occurs when $Q = 0$ and the QP reduces to a linear program. The code is written to take maximum advantage of this situation, so that it also provides an efficient method for linear programming.

Hardware/software environment

BQPD is a single-precision Fortran 77 subroutine, and parameters are passed through arguments. Matrix information is processed in auxiliary files, and alternative modules can be plugged in to provide either a dense or a sparse format for representing and manipulating the basis matrix.

Contact address

>Roger Fletcher
>Department of Mathematics and Computer Science
>University of Dundee
>Dundee DD1 4HN
>Scotland, United Kingdom

Additional comments

Special features include a constraint prescaling routine and full documentation through comments in the code. Special arrangements can be made for use in a commercial environment.

Reference

R. Fletcher, *Resolving Degeneracy in Quadratic Programming*, Numerical Analysis Report NA/135, Department of Mathematics and Computer Science, University of Dundee, Dundee, Scotland, UK, 1991.

BT

Areas covered by the software

Minimization of locally Lipschitz functions subject to box constraints, as well as linear equality and affine inequality constraints

Basic algorithms

All algorithms are based on a combination of the bundle concept developed for the minimization of locally Lipschitz, weakly semismooth functions and the trust-region philosophy, now widely used in optimization. For the solution of inner quadratic programming problems, an algorithm due to Powell is used.

Hardware/software environment

Software is written in Fortran 77 in double precision. The user must provide a common block /CMACHE/EPS in subroutine QL0001, where EPS defines the underlying machine precision. The user must also provide a subroutine for computing the objective function and an arbitrary subgradient at any admissible point.

Contact addresses

Both Jochem Zowe and Jiri V. Outrata can be reached at the following address:

> Mathematical Institute
> University of Bayreuth
> Mailbox 10 12 51
> W-8580 Bayreuth
> Germany
> jochem.zowe@uni-bayreuth.dbp.de
> jiri.outrata@uni-bayreuth.dbp.de

References

C. Lemaréchal, J.-J. Strodiot, and A. Bihain, *On a bundle algorithm for nonsmooth optimization*, in Nonlinear Programming 4, O. L. Mangasarian, R. R. Meyer, and S. M. Robinson, eds., Academic Press, New York, 1991.

J. Outrata, J. Zowe, and H. Schramm, *Bundle Trust Methods: Fortran Codes for Nondifferentiable Optimization*, User's Guide, DFG Report No. 269, 1991.

H. Schramm and J. Zowe, *A version of the bundle idea for minimizing a nonsmooth function: Conceptual idea, convergence analysis, numerical results*, SIAM J. Optim., 2 (1992), pp. 121–152.

BTN

Areas covered by the software

Unconstrained minimization in a parallel computing environment. The software is especially suited to problems with large numbers of variables.

Basic algorithms

BTN uses a block, truncated Newton method based on a line search. The block conjugate gradient method is used to compute an approximation to the Newton direction by approximately solving the Newton equations. The blocking is used to achieve parallelism in both the linear algebra and the function evaluations.

Both easy-to-use and customized versions are provided. The easy-to-use version requires only that the user provide a (scalar) subroutine to evaluate the objective function and its gradient of first derivatives; no knowledge of parallel computing is required. The customized version allows more complicated usage; for example, it allows the function to be evaluated in parallel.

A parallel derivative checker is also included in the package. The software can be run on traditional computers to simulate a parallel computing environment.

Hardware/software environment

The software is written in ANSI Fortran 77, with a small number of machine-dependent subroutine calls and compiler directives to control the parallel execution. Machine constants are set in a single subroutine (d1mach). Versions of the software are available for the Intel iPSC/2 and iPSC/860 hypercube computers (distributed memory) and the Sequent Balance and Symmetry parallel computers (shared memory). The Sequent version can also be run on traditional computers, where it will simulate the performance of a parallel machine. The software is currently available only in double precision.

Contact address

>Stephen G. Nash
>ORAS Department
>George Mason University
>Fairfax, VA 22030
>Phone: (703) 993-1678
>snash@gmuvax.gmu.edu

References

S. G. Nash and A. Sofer, *BTN: Software for parallel unconstrained optimization*, ACM Trans. Math. Software, 18 (1992), pp. 414–448.

——, *A general-purpose parallel algorithm for unconstrained optimization*, SIAM J. Optim., 1 (1991), pp. 530–547.

SOFTWARE PACKAGES

CNM

Areas covered by the software

Low storage algorithms in Pascal for linear algebra and minimization problems. The scope of the package is described in the book *Compact Numerical Methods for Computers: Linear Algebra and Function Minimisation*, which is cited below.

Basic algorithms

There are 27 algorithms supported by over 50 other files of driver and example code. The function minimizers include Hooke and Jeeves, Nelder–Mead, conjugate gradient (choice of three update formulas), variable metric (modified Fletcher 1970 code), and a modified Marquardt nonlinear least squares method.

The commented algorithms are in book form, with a coupon for the diskette. The codes are also in the Pascal directory of *netlib*.

Hardware/software environment

The software is intended to run in Turbo Pascal, versions 3.01a or 5.x primarily on IBM and IBM-compatible PCs. A BATch command file is included to run a complete example test of all the routines, saving the results in output files that are images of the screens.

Contact address

> American Institute of Physics
> Marketing Services
> 335 East 45th Street
> New York, NY 10017-3483
> Phone: (800) 445-6638

Reference

J. C. Nash, *Compact Numerical Methods for Computers: Linear Algebra and Function Minimisation*, 2nd ed., Adam Hilger Ltd. (Institute of Physics Publications), Bristol, England, UK, 1990.

CONOPT

Areas covered by the software

General nonlinear programming models with sparse nonlinear constraints and nonlinear objective functions

Basic algorithms

The algorithm in CONOPT is based on the generalized reduced-gradient (GRG) algorithm. All matrix operations are implemented by using sparse matrix techniques to allow very large models. Without compromising the reliability of the GRG approach, the overhead of the GRG algorithm is minimized by, for example, using dynamic feasibility tolerances, reusing Jacobians whenever possible, and using an efficient reinversion routine. The algorithm uses many dynamically set tolerances and therefore runs, in most cases, with default parameters.

Hardware/software environment

CONOPT is available in a stand-alone version, a callable library, and a subsystem under GAMS. All versions are distributed in compiled form. CONOPT is available for PCs, most workstations, and many mainframes.

Contact address

For general information, contact
> ARKI Consulting and Development A/S
> Bagsvaerdvej 246 A
> DK-2880 Bagsvaerd
> Denmark
> Phone: +45 44 49 03 23
> Fax: +45 44 49 03 33

For information about GAMS/CONOPT in the United States, contact
> GAMS Development Corp.
> 1217 Potomac Street NW
> Washington, DC 20007
> Phone: (202) 232-5662
> Fax: (202) 483-1921

Additional comments

The system is continually updated, mainly to improve reliability and efficiency on large models.

Reference

A. S. Drud, *CONOPT—A large scale GRG code*, ORSA J. Comput., to appear.

CONSOL-OPTCAD

Areas covered by the software

Optimization-based design of engineering systems. Allows multiple objectives and soft and hard constraints. Handles functional (semi-infinite) specifications. Interactive tradeoff exploration. Easily coupled with user-supplied simulators; interfaces with MATLAB, SIMNON, WATAND, SPICE are available from the authors.

Basic algorithms

The given problem is transformed into a constrained, weighted minimax problem, with a different transformation depending on whether the soft/hard constraints are satisfied. The minimax problem is solved by using feasible sequential quadratic programming (see the entry on FSQP), extended to accommodate semi-infinite constraints. The user interactively tightens or relaxes specifications by adjusting "good values" and "bad values" (good/bad "curves" in the case of functional specifications), which determine the weights in the minimax problems.

Hardware/software environment

Software runs only on Sun computers. It makes use of dynamic loading.

Contact addresses

André L. Tits
Electrical Engineering Dept.
and Systems Research Center
University of Maryland
College Park, MD 20742
Phone: (301) 405-3669
Fax: (301) 405-6707
andre@src.umd.edu

Carolyn Garrett
Office of Technology Liaison
4312 Knox Road
University of Maryland
College Park, MD 20742
Phone: (301) 405-4209
Fax: (301) 314-9871
cg54@umail.umd.edu

Additional comments

A detailed manual is available. The package is available free of charge to academic institutions with ftp access. An elaborate graphical user interface with graphical input capability is under development.

References

M. K. H. Fan, A. L. Tits, J. Zhou, L.-S. Wang, and J. Koninckx, *CONSOLE User's Manual, Version 1.1*, TR 87-212r1, Systems Research Center, University of Maryland, College Park, MD, 1990.

M. K. H. Fan, L.-S. Wang, J. Koninckx, and A. L. Tits, *Software package for optimization-based design with user-supplied simulators*, IEEE Control Systems Magazine, 9 (1989), pp. 66–71.

CPLEX

Areas covered by the software

Ready-to-run applications:

    CPLEX Linear Optimizer

    CPLEX Mixed-Integer Optimizer

Optimization libraries callable from C, Fortran, and Pascal programs:

    CPLEX Callable Library

    CPLEX Mixed-Integer Library

The CPLEX Linear Optimizer and Callable Library solve linear programming problems. The CPLEX Mixed-Integer Optimizer and Mixed-Integer Library solve integer programming problems as well as linear programming problems. Both linear and integer packages also solve network-structured problems with unlimited side constraints.

CPLEX products are designed to solve large, difficult problems where other linear programming solvers fail or are unacceptably slow. CPLEX algorithms are exceptionally fast and robust, providing exceptional reliability, even for poorly scaled or numerically difficult problems.

Typical areas of application include large models in refining, manufacturing, banking, finance, transportation, timber, defense, energy, and logistics. CPLEX is also used heavily in academic research in universities throughout the world.

Basic algorithms

The CPLEX linear programming packages use a "modified primal and dual simplex" algorithm with multiple algorithm options for pricing and factorization. An optional preprocessor is available for problem reduction. Most algorithmic parameters can be manually adjusted by users, although preset defaults with built-in dynamic adjustment often provide the best performance.

The CPLEX mixed-integer programming systems use branch-and-bound and branch-and-cut algorithms. Several options are available to guide and limit integer solving, such as user-set priorities and alternative branching and node-selection strategies. Specialized algorithms for specially ordered sets are available for use on certain classes of problem.

All CPLEX products include a special algorithm for extracting and solving a maximal size network from within linear programming problems. The network algorithm is much more efficient than LP algorithms where pure network structures can be identified.

Hardware/software environment

All CPLEX products are designed to be portable and are available on most popular hardware/software environments, including personal computers (all IBM compatibles); UNIX workstations (Sun, HP/Apollo, DEC, IBM, MIPS, and others); mainframes (IBM, DEC, Unisys, and others); and supercomputers (Cray, Convex).

Contact address

    CPLEX Optimization, Inc.
    Suite 279
    930 Tahoe Blvd., Bldg. 802
    Incline Village, NV 89451
    Phone: (702) 831-7744
    Fax: (702) 831-7755

Additional comments

CPLEX solvers are available in two forms:

> The CPLEX Linear Optimizer and Mixed-Integer Optimizer are complete applications designed for ease of use. Because a complete on-line help system exists, most users never open the documentation and are running problems within minutes of opening the package.

> The CPLEX Callable Library and Mixed-Integer Library are in the form of callable routines that can be used to embed optimization functionality within user-written applications. The callable products were designed to simplify development while providing the flexibility developers require. CPLEX offers a Value-Added Reseller Program for developers interested in using CPLEX within their own products.

CPLEX reads linear and integer problems in several formats, including MPS format and CPLEX LP format. CPLEX also interfaces with several modeling languages, including GAMS, AMPL, and MPL.

To support academic research, CPLEX is offered at significant discounts to academic institutions.

References

R. E. Bixby, *Implementing the simplex method: The initial basis*, ORSA J. Comput., 4 (1992), pp. 267–284.

M. Burgard, *Solutions to Complex Problems*, UNIX World, 7 (January 1990), pp. 111–113.

R. Barr, *A large-scale linear optimizer for small- and large-scale systems*, OR/MS Today, 18 (August 1991), pp. 26–29.

C-WHIZ

Areas covered by the software
Linear programming models

Basic algorithms
C-WHIZ is an in-core implementation of the simplex algorithm, supplemented with extensive pre- and postsolve procedures to simplify the matrix structure, and a crashing procedure to arrive quickly at an advanced starting basis. In addition to a fast primal algorithm, C-WHIZ contains a stable primal algorithm, which is automatically invoked for numerically difficult matrices, and a dual algorithm, which is normally used in the solution of mixed-integer models. C-WHIZ can solve matrices of up to 32,000 rows and a virtually unlimited number of columns, taking advantage of matrix sparsity to make economical use of available memory. C-WHIZ accepts matrix input in standard MPS format or from the MPSIII database.

Hardware/software environment
C-WHIZ is written in ANSI C and is currently available on PCs under Extended DOS, as well as on the IBM RS/6000, Sun, and HP 700 series UNIX workstations. An assembly language version of C-WHIZ is available on IBM and IBM-compatible mainframes.

Contact addresses

Ketron Management Science
1700 N. Moore St.
Suite 1710
Arlington, VA 22209
Phone: (703) 558-8701

R. Staudinger
ARBED Information Systems
19, Avenue de la Liberté
Luxembourg
Phone: +352 4792-2107

Additional comments
C-WHIZ is undergoing continual enhancement. A version that removes the 32,000 row limit is being prepared.

Reference
C-WHIZ User's Manual, Ketron Management Science, Arlington, VA, December 1991.

DFNLP

Areas covered by the software
DFNLP solves constrained nonlinear data fitting and minimax problems, where the objective function may be a sum of squares of function values, a sum

of absolute function values, a maximum of absolute function values, or a maximum of functions. In addition, there may be any set of equality or inequality constraints. It is assumed that all individual problem functions are continuously differentiable.

Basic algorithms

By introducing additional variables and constraints, the problem is transformed into a general, smooth, nonlinear programming problem, which is then solved by NLPQL. For least squares problems, typical features of special-purpose algorithms are retained (i.e., a combination of a Gauss–Newton and a quasi-Newton search direction). In this case, the additional introduced variables are eliminated in the quadratic programming subproblem, so that calculation time is not increased significantly.

Hardware/software environment

DFNLP is a double-precision Fortran 77 subroutine, and parameters are passed through arguments.

Contact address

> Klaus Schittkowski
> Mathematisches Institut
> Universität Bayreuth
> 8580 Bayreuth, Germany
> Klaus.Schittkowski@uni-bayreuth.de

Additional comments

Special features include reverse communication, internal scaling, feasibility with respect to bounds and linear constraints, and documentation in the form of comments within the code.

Reference

K. Schittkowski, *Solving constrained nonlinear least squares problems by a general purpose SQP-method*, in Trends in Mathematical Optimization, K.-H. Hoffmann, J.-B. Hiriart-Urruty, C. Lemaréchal, and J. Zowe, eds., International Series of Numerical Mathematics, Vol. 84, Birkhäuser-Verlag, Basel, Switzerland, 1985.

DOC—Design Optimization Control Program

Areas covered by the software

Nonlinear programming, multiobjective and discrete variable optimization

Basic algorithms

DOC is a control program that can be used in conjunction with DOT (see separate entry) and user-supplied subroutines to produce a powerful customized optimization program. Once this program has been created, the user can change the mix of constraints, the weighting of objectives, the output format, etc., simply by changing a few input parameters to DOC.

Hardware/software environment

Software is written in ANSI Fortran.

Contact address

>VMA Engineering
>225 S. Cheyenne Mountain Blvd.
>Suite 200B
>Colorado Springs, CO 80806
>Phone: (719) 527-2691
>Fax: (719) 527-2692

Additional comments

Additional features of DOC include parametric studies and optimization based on curve fits. The data-fitting capabilities are useful for optimization based on the results of physical experiments.

Hotline support is provided.

References

DOC User's Manual, VMA Engineering, Goleta, CA, 1993.

G. N. Vanderplaats, *Numerical Optimization Techniques for Engineering Design, with Applications*, McGraw–Hill, New York, 1984.

DOT—Design Optimization Tools

Areas covered by the software

Nonlinear programming

Basic algorithms

The DOT program uses the BFGS method for unconstrained optimization problems. For constrained problems, DOT implements a modified feasible directions algorithm (similar to a reduced-gradient method), sequential linear programming, and sequential quadratic programming. Gradients can either be supplied by the user or obtained by finite differencing. A line search is performed along each search direction generated by one of the methods.

Hardware/software environment
Software is written in ANSI Fortran.

Contact address

> VMA Engineering
> 225 S. Cheyenne Mountain Blvd.
> Suite 200B
> Colorado Springs, CO 80806
> Phone: (719) 527-2691
> Fax: (719) 527-2692

Additional comments
DOT is the latest software written by the author of the CONMIN and ADS programs. DOT is fully supported and is used worldwide.

Hotline support is provided.

References

DOT User's Manual, VMA Engineering, Goleta, CA, 1992.

G. N. Vanderplaats, *Numerical Optimization Techniques for Engineering Design, with Applications*, McGraw–Hill, New York, 1984.

FortLP

Areas covered by the software
Linear and mixed-integer linear programming problems

Basic algorithms
The present version of FortLP uses a sparse implementation of the revised simplex method for linear programming. A branch-and-bound technique is used to obtain integer solutions. The basis matrix is maintained in sparse LU form and is updated by the Forrest–Tomlin update procedure.

Hardware/software environment
The software is written in ANSI standard Fortran 77. The current version uses INTEGER*2 constructs to preserve space, but these are likely to be removed at the next release, in light of increasing availability of storage. These nonstandard constructs may impose some restrictions on portability; otherwise, FortLP is expected to run on most machines.

FortLP accepts problem data in MPSX format but is written in a modular way to allow users to use their own methods for supplying input.

Contact addresses

NAG, Inc.
1400 Opus Place
Suite 200
Downers Grove, IL 60515-5702
Phone: (708) 971-2337
Fax: (708) 971-2706

NAG Ltd.
Wilkinson House, Jordan Hill Road
Oxford OX2 8DR
England, United Kingdom
Phone: +44 (865) 511245
Fax: +44 (865) 310139

Additional comments

Interior-point methods with a crossover to the simplex method during a final phase are being developed.

The package is being distributed as source code. It is fully supported and maintained, and regular updates are produced.

FSQP

Areas covered by the software

Nonlinear standard (non-minimax) or minimax optimization problems with general inequality constraints and linear equality constraints

Basic algorithms

Algorithms in FSQP are based on the concept of feasible sequential quadratic programming. Starting with a feasible point, provided by the user or generated automatically, these algorithms produce successive iterates that all satisfy the constraints. After feasibility has been reached, the objective function can be decreased either after each iteration with an Armijo-type arc search or after at most three iterations with a nonmonotone line search. The user has the option to choose one of the two searches. The merit function used in both searches is the objective function itself. Both algorithms in FSQP can be shown to have global and two-step superlinear convergence properties. The user can provide routines for gradients of objective and constraint functions or require that gradients be computed by finite differences.

Hardware/software environment

Software is written in ANSI Fortran 77. A routine that computes the machine precision is included.

Contact address

André L. Tits
Electrical Engineering Dept. and Systems Research Center
University of Maryland
College Park, MD 20742

SOFTWARE PACKAGES

Phone: (301) 405-3669
Fax: (301) 405-6707
`andre@src.umd.edu`

Additional comments

The package includes test problems and a detailed user's guide. It is available free of charge to academic institutions and government laboratories. The current version number is 2.4b.

References

J. F. Bonnans, E. R. Panier, A. L. Tits, and J. L. Zhou, *Avoiding the Maratos effect by means of a nonmonotone line search. II: Inequality constrained problems—Feasible iterates*, SIAM J. Numer. Anal., 29 (1992), pp. 1187–1202.

E. R. Panier and A.L. Tits, *On combining feasibility, descent and superlinear convergence in inequality constrained optimization*, Math. Programming, 59 (1993), pp. 261–276.

J. L. Zhou and A. L. Tits, *Nonmonotone line search for minimax problems*, J. Optim. Theory Appl., 76 (1993), pp. 455–476.

———, *User's Guide for FSQP Version 2.4—A Fortran Code for Solving Optimization Problems, Possibly Minimax, with General Inequality Constraints and Linear Equality Constraints, Generating Feasible Iterates*, SRC TR-90-60r1c, Systems Research Center, University of Maryland, College Park, MD, 1991.

GAMS

Areas covered by the software

Linear programming, nonlinear programming, mixed-integer optimization

Basic algorithms

GAMS (General Algebraic Modeling System) was developed by Meeraus and Brooke of the GAMS Development Corporation. For linear and nonlinear programming, GAMS provides an interface to the MINOS system developed by Murtagh and Saunders (see separate entry). For mixed-integer programming, it interfaces with ZOOM-XMP, which was developed by Marsten.

Hardware/software environment

GAMS is available in a number of formats that have different platform requirements.

The GAMS Student Edition consists of a copy of the 280-page *User's Guide* and a student edition of the GAMS software, which runs on an IBM or IBM-compatible PC and is supplied on either 3.5-inch or 5.25-inch disks. For the MINOS module, there are space limitations of 500 nonzero elements in the constraint matrix, 100 of which can be "nonlinear nonzeros." For the ZOOM-XMP module, there is a limit of 20 discrete variables.

GAMS 2.25 is available for a number of environments, including DOS-compatible PCs with 640KB RAM (math coprocessor recommended), DOS-compatible 80386 (with 80187/387), or DOS-compatible 80486 computers (at least 2 MB of memory recommended); workstations (Sun4/Sparc, IBM RS/6000, DEC MicroVax, DEC stations, Hewlett–Packard HP 9000, 300, 700, and 800 series, Apollo 3000, 4000, 10000 series, SGI); and mainframes (IBM under VM/CMS, DEC Vax under VMS).

Limitations on problem size include a limit of 32,767 rows in the constraint matrix for the MINOS module. Additional modules (which link GAMS to other numerical optimization codes) are available as options. Some of these are available only for certain environments. These include MPSX, OSL, and XA (for linear programming or mixed-integer programming); and CONOPT and DICOPT (for nonlinear mixed-integer programming).

Contact address
>The Scientific Press
651 Gateway Blvd., Suite 1100
South San Francisco, CA 94080-7014
Phone: (800) 451-5409 (outside California)
Phone: (415) 583-8840 (inside California)
Fax: (415) 583-6371

Additional comments

GAMS is a high-level modeling language that is formally similar to Fortran. It allows the user to describe models in concise algebraic statements that are readily comprehensible to other readers. Features such as sensitivity analysis and report generation are provided.

Reference

GAMS: A User's Guide, The Scientific Press, San Francisco, CA, 1988.

GAUSS

Areas covered by the software

The software consists of nine packages that are designed for use with the GAUSS matrix programming language. Four of these packages relate to optimization; the other five relate to statistical analysis. The four relevant packages are

OPTIMUM	Unconstrained optimization
MAXLIK	Maximum likelihood estimation
NLSYS	Solving nonlinear equations
SIMPLEX	Linear programming

Basic algorithms

The OPTIMUM module contains procedures for unconstrained minimization. The user may select from several options: Newton's method, quasi-Newton methods (DFP, BFGS, and scaled BFGS), steepest-descent, and Polak–Ribiere conjugate gradient (PRCG). Four line-search algorithms are available: unit step length, golden section line search, a procedure due to Brent, and a procedure that uses a cubic or quadratic model of the objective function along the search direction.

The user must supply procedures for computing the function and has the option of providing procedures for computing the gradient, the Hessian, or both.

Several run-time switches are provided so that the user may change the algorithm choice or step-length routine during the execution of the module, and hence control the convergence properties interactively.

The SIMPLEX module contains procedures for solving linear programming problems. It uses the two-phase standard revised simplex method with a factorization similar to the product form of the inverse.

Hardware/software environment

All of the GAUSS applications modules are written in GAUSS programming language. Source code is provided. The software requires an IBM-compatible machine running DOS, with an Intel-compatible math coprocessor, a minimum of 4 MB of RAM, and a hard disk with at least 4 MB of available space.

Contact address

Aptech Systems, Inc.
23804 SE Kent–Kangley Road
Maple Valley, WA 98038
Phone: (206) 432-7855
Fax: (206) 432-7832

Additional comments

These packages are being revised and expanded.

GENESIS—Structural Optimization Software

Areas covered by the software

Finite element analysis and optimization of large-scale structures

Basic algorithms

GENESIS is a fully integrated analysis and optimization program, written from the start as a design program. The finite element method is used for statics and normal modes analysis. The analysis data is a subset of the well-known NASTRAN bulk data. Design optimization is performed by using the latest approximation methods, many of which were developed by VMA personnel. A high-quality approximation based on a full analysis is created and solved with the DOT optimizer. The process is repeated until convergence to an optimum is achieved. Typically, fewer than ten detailed analyses are required.

GENESIS performs both sizing (member dimension) and shape (geometry) optimization. Typical problem sizes involve well over 100 design variables and many thousands of constraints. Almost any calculated response (or nonlinear function of variables and responses) can be chosen as the objective or may be constrained. These include mass, volume, eigenvalues, stresses, and deflections. Pre- and postprocessing is done by the PDA/PATRAN and SDRC/IDEAS programs.

Hardware/software environment

Software is written in ANSI Fortran. There are under 30 machine-dependent lines of code. GENESIS is operational on a wide range of platforms, from workstations through supercomputers.

Contact address

VMA Engineering
225 S. Cheyenne Mountain Blvd.
Suite 200B
Colorado Springs, CO 80806
Phone: (719) 527-2691
Fax: (719) 527-2692

Additional comments

GENESIS is available directly from VMA Engineering and also from PDA Engineering as the P3/STRUCTURAL OPTIMIZATION module in PATRAN.
    Hotline support is provided.

References

GENESIS User's Manuals, VMA Engineering, Goleta, CA, 1993.

G. N. Vanderplaats, *Thirty years of modern structural optimization*, Adv. Engrg. Software, 16 (1993), pp. 81–88.

GENOS 1.0

Areas covered by the software
Nonlinear optimization for problems with network and generalized network constraints

Basic algorithms
The algorithms in GENOS 1.0 include network simplex, primal truncated Newton, and simplicial decomposition.

Hardware/software environment
Software is written in ANSI Fortran, in double precision. Machine dependencies are restricted to the subroutine SETMC that defines machine-dependent constants. Versions of the code that are optimized for vector and parallel architectures are available.

Contact addresses

John Mulvey
Department of Civil Engineering
Princeton University
Princeton, NJ 08544
mulvey@macbeth.princeton.edu

Stavros A. Zenios
The Wharton School
University of Pennsylvania
Philadelphia, PA 19104
zenios@wharton.upenn.edu

References

R. S. Dembo, J. M. Mulvey, and S. A. Zenios, *Large-scale nonlinear network models and their application*, Oper. Res., 37 (1989), pp. 353–372.

J. M. Mulvey and S. A. Zenios, *GENOS 1.0 User's Guide: A Generalized Network Optimization System*, Report 87-13-03, Department of Decision Science, The Wharton School, University of Pennsylvania, Philadelphia, PA, 1987.

GINO

Areas covered by the software
Constrained and unconstrained optimization, linear and nonlinear equations, systems of inequalities. Contains a built-in library of trigonometric, probability, financial, and general mathematical functions.

Basic algorithms
Lasdon and Waren's GRG2 nonlinear programming algorithm is the core algorithm. The LINDO interactive interface is used.

Hardware/software environment

GINO is available in a number of formats that have different platform requirements. These are summarized in the table below.

	GINO	Super GINO	Hyper GINO	Industrial GINO
Variables	50	100	200	800
Constraints	30	50	100	400
Memory—PC	256KB	512KB	1MB	1MB
Memory—Macintosh	512KB	512KB	1MB	4MB
Memory—Workstation	n.s.	n.s.	n.s.	4MB

n.s. = not supported.

Mainframe versions are also available. Workstations supported are DEC 3100, DEC 5000, HP, IBM RS/6000, MicroVAX, NeXT, Sun SPARCstation, and Sun3.

Contact address

LINDO Systems, Inc.
P. O. Box 148231
Chicago, IL 60614
Phone: (800) 441-BEST and (312) 871-2524
Fax: (312) 871-1777

Reference

J. Liebman, L. Lasdon, L. Schrage, and A. Waren, *Modeling and Optimization with GINO*, The Scientific Press, San Francisco, CA, 1986.

GRG2

Areas covered by the software

Nonlinear programming

Basic algorithms

GRG2 uses an implementation of the generalized reduced gradient (GRG) algorithm. It seeks a feasible solution first (if one is not provided) and then retains feasibility as the objective is improved. It uses a robust implementation of the BFGS quasi-Newton algorithm as its default choice for determining a search direction. A limited-memory conjugate gradient method is also available, permitting solutions of problems with hundreds or thousands of variables. The problem Jacobian is stored and manipulated as a dense matrix, so the effective size limit is one to two hundred active constraints (excluding simple bounds on the variables, which are handled implicitly).

The GRG2 software may be used as a stand-alone system or called as a subroutine. The user is not required to supply code for first partial derivatives of problem functions; forward or central difference approximations may be used instead. Documentation includes a 60-page user's guide, in-line documentation for the subroutine interface, and complete installation instructions.

Hardware/software environment

GRG2 is written in ANSI Fortran. A C version is also available. Machine dependencies are relegated to the subroutine INITLZ, which defines three machine-dependent constants.

Contact address

> Leon Lasdon
> MSIS Department
> College of Business Administration
> University of Texas
> Austin, TX 78712-1175
> Phone: (512) 471-9433

Additional comments

GRG2 is by far the most widely distributed nonlinear programming code. It is the optimizer in the "Solver" option of the popular Excel 3.0 spreadsheet. A similar facility called "What-If Solver" is available as an add-on release for release 2.X of Lotus 1-2-3 (contact Frontline Systems, Inc. at (415) 329-6877). GRG2 is also available as the nonprocedural interactive package GINO.

Reference

L. S. Lasdon, A. D. Waren, A. Jain, and M. Ratner, *Design and testing of a generalized reduced gradient code for nonlinear programming*, ACM Trans. Math. Software, 4 (1978), pp. 34–50.

HOMPACK

Areas covered by the software

Systems of nonlinear equations and polynomial systems

Basic algorithms

The algorithms in HOMPACK are based on probability-one homotopy maps and are globally convergent with probability one under mild assumptions. There are three qualitatively different algorithms (ODE based, normal flow, augmented Jacobian matrix) for tracking the homotopy zero curve, and also different codes for dense and sparse problems. The code is modular and arranged hierarchically, so the user can (1) call a driver supplying very little information or (2) call the tracking routines directly with complete control.

HOMPACK includes special algorithms for finding *all* solutions to a polynomial system, with a simple tableau input format for the polynomial coefficients. The routines are self-documenting; test programs are included.

Hardware/software environment
The code is written in double-precision ANSI Fortran 77. Machine dependencies are restricted to the subroutine D1MACH.

Contact address
Software can be obtained from *netlib* (`send index from hompack`) or from

>Layne T. Watson
>Department of Computer Science
>Virginia Polytechnic Institute & State University
>Blacksburg, VA 24061-0106
>Phone: (703) 231-7540
>`ltw@vtopus.cs.vt.edu`

Additional comments
The sparse subroutines are currently under revision.

References

A. P. Morgan, A. J. Sommese, and L. T. Watson, *Finding all isolated solutions to polynomial systems using HOMPACK*, ACM Trans. Math. Software, 15 (1989), pp. 93–122.

L. T. Watson, *Numerical linear algebra aspects of globally convergent homotopy methods*, SIAM Rev., 28 (1986), pp. 529–545.

L. T. Watson, S. C. Billups, and A. P. Morgan, *Algorithm 652: HOMPACK: A suite of codes for globally convergent homotopy algorithms*, ACM Trans. Math. Software, 13 (1987), pp. 281–310.

IMSL Fortran and C Library

Areas covered by the software
Figures 1 and 2 provide a quick reference to the optimization routines of Chapter 8 of the Fortran Library. Only part of this coverage is available for the C Library. There are also routines for solving systems of nonlinear equations in Chapter 7 of both products.

SOFTWARE PACKAGES

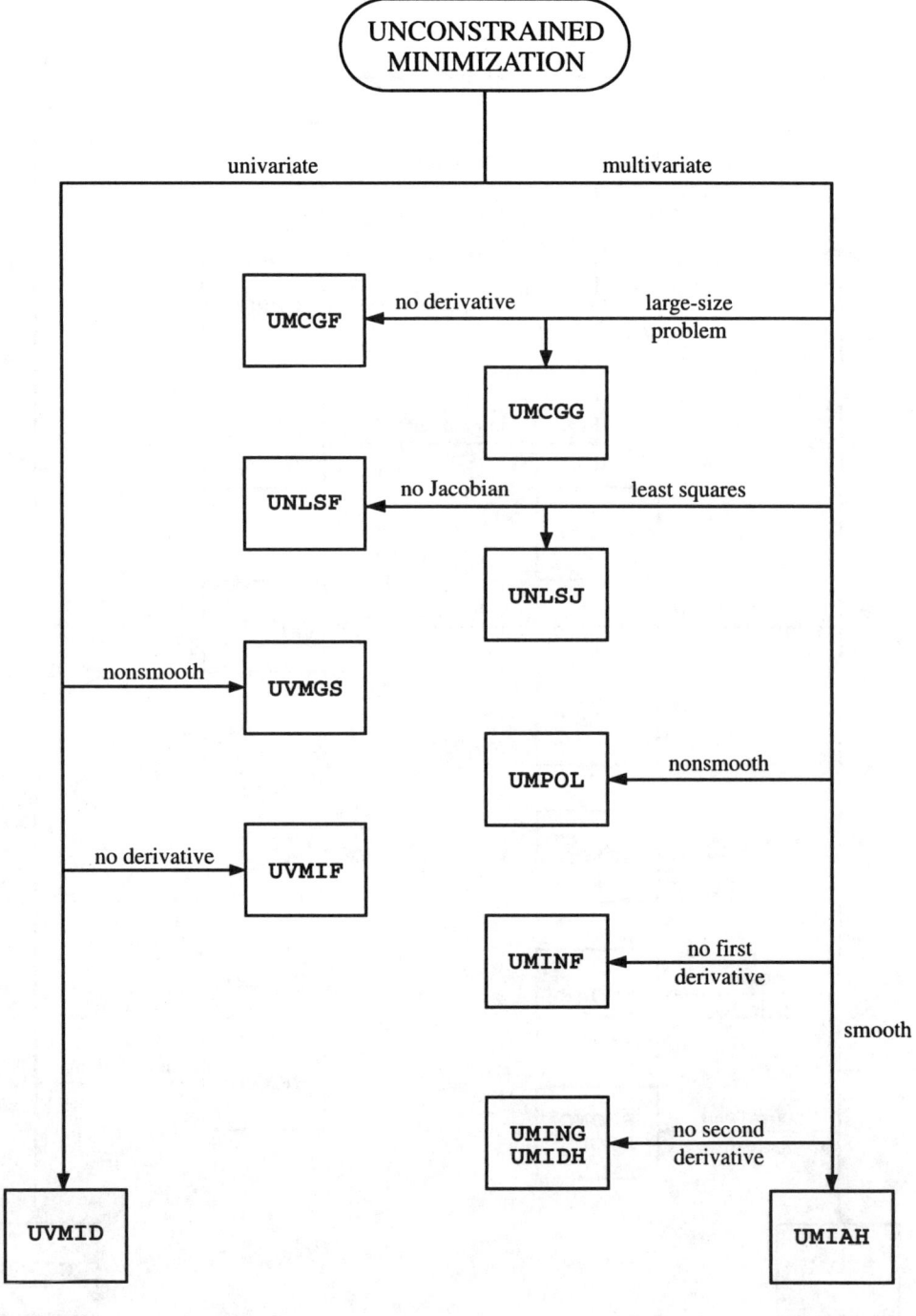

FIG. 1. *Unconstrained minimization subroutines in the IMSL library.*

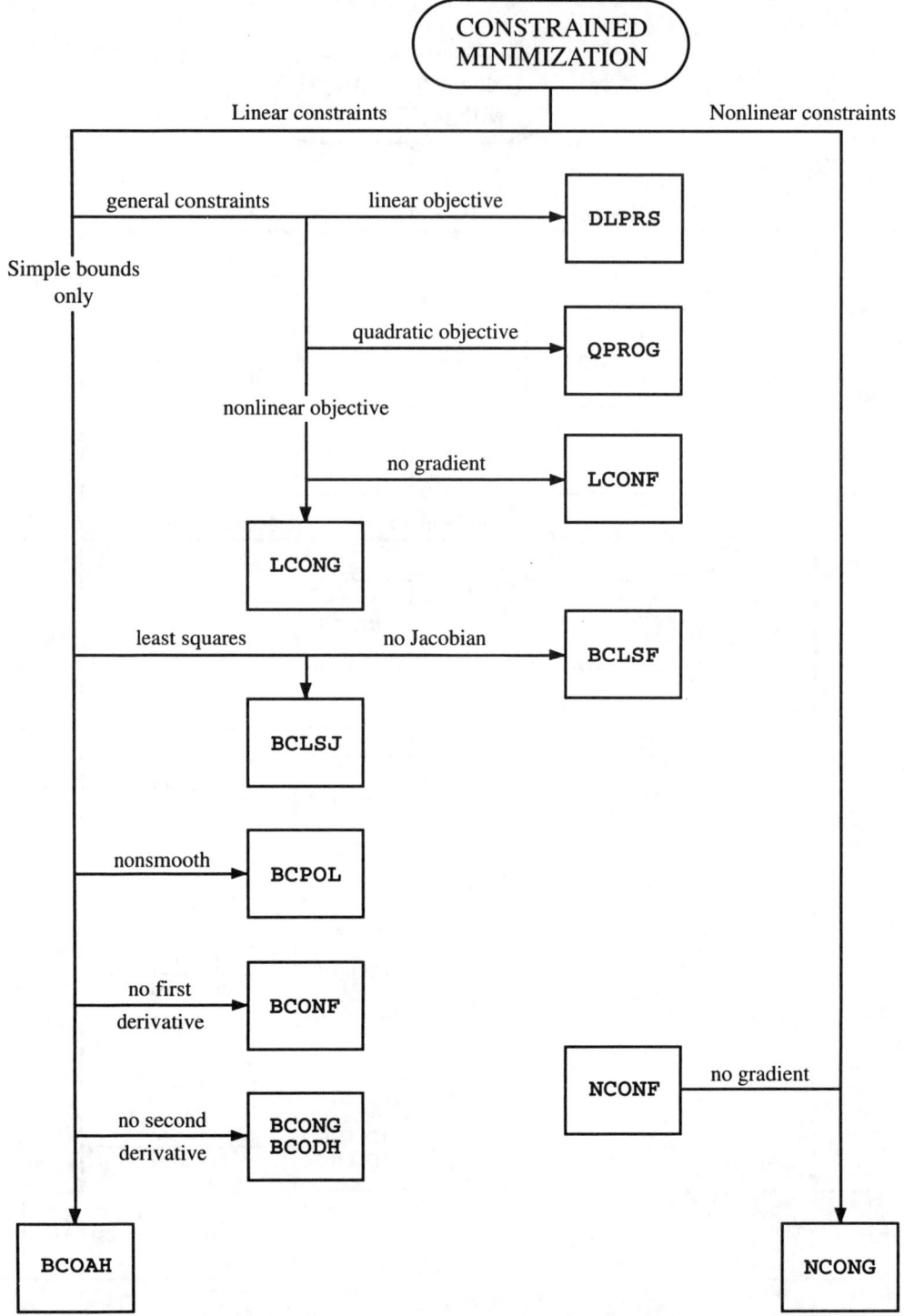

FIG. 2. *Constrained minimization subroutines in the IMSL library.*

Basic algorithms

The Fortran and C Library manuals discuss the algorithms, with references as well as sample programs.

Hardware/software environment

The Fortran Library is written in ANSI standard Fortran 77 and can be implemented on a wide variety of machines, including workstations. The C Library is written in ANSI C. It is available on several workstations.

Contact address

This software and other related products can be obtained by licensing through

> IMSL, Inc.
> One Sugar Creek Place
> 14141 Southwest Freeway
> Sugar Land, TX 77478
> Phone: (713) 242-6776
> Fax: (713) 242-9799

Additional comments

Users can include the software in their derivative works with a licensing agreement. Contact IMSL, Inc., Sales, for more details.

LAMPS

Areas covered by the software

Linear programming and mixed-integer programming

Basic algorithms

LAMPS (Linear and Mathematical Programming System) offers a primal and a dual simplex algorithm for the solution of linear programs, and a branch-and-bound algorithm for mixed-integer programs. The linear programming method used for solution of nodes in the branch-and-bound tree is user selectable.

LAMPS also provides sensitivity analysis in terms of parametric variation of the right-hand side or objective function, and ranging.

Hardware/software environment

Software is normally supplied only in object code and executable form. Versions are available for 80386/486 MS-DOS machines, most Unix workstations, VAX, Prime, Convex, and Cray. Algorithms and data/matrix manipulation functions of LAMPS are available as a subroutine library.

Contact addresses

Advanced Mathematical Software, Inc., at either

4 Yukon Road	186 North Elm Street
London, SW12 9PU	Northhampton, MA 01060
England, United Kingdom	
Phone: +44 (081) 675-4766	Phone: (413) 584-1605
Fax: +44 (081) 675-4880	Fax: (413) 586-2501

Additional comments

LAMPS is designed for the solution of large problems, although it will operate efficiently on small- and medium-sized problems. Most standard input formats are acceptable, and output (solution) reporting is very flexible.

Mixed-integer problems may define data in terms of S1 or S2 sets, general integer, binary, and semicontinuous variables.

Algorithms of LAMPS are also available for direct use with MAGIC (a matrix generation and reporting system) and GAMS.

Free testing and advice are always available.

LANCELOT (Release A)

Areas covered by the software

Unconstrained optimization problems, systems of nonlinear equations, nonlinear least squares, bound-constrained optimization problems, and general nonlinearly constrained optimization problems. Special emphasis is placed on large-scale problems.

Basic algorithms

The LANCELOT package uses an augmented Lagrangian approach to handle all constraints other than simple bounds. The bounds are dealt with explicitly at the level of an outer-iteration subproblem, where a bound-constrained nonlinear optimization problem is approximately solved at each iteration.

The algorithm for solving the bounded problem combines a trust-region approach adapted to handle the bound constraints, projected gradient techniques, and special data structures to exploit the (group partially separable) structure of the underlying problem.

The software additionally provides direct and iterative linear solvers (for Newton equations), a variety of preconditioning and scaling algorithms for more difficult problems, quasi-Newton and Newton methods, provision for analytical and finite-difference gradients, and an automatic decoder capable of reading problems expressed in Standard Input Format (SIF).

Hardware/software environment

LANCELOT A is written in standard ANSI Fortran 77. Single- and double-precision versions are available. Machine dependencies are isolated and easily adaptable. Automatic installation procedures are available for DEC VMS, DEC ULTRIX, Sun UNIX, Cray UNICOS, IBM VM/CMS, and IBM AIX.

Contact address

A low-cost version of the software is available for academic and research institutions, under special conditions. For such a version, contact

> Philippe Toint
> Department of Mathematics
> FUNDP
> 61 rue de Bruxelles
> B5000 Namur, Belgium
> pht@math.fundp.ac.be

Commercial licenses for the software are subject to negotiations between the interested party and one of the authors. There are two other authors in addition to Philippe Toint:

Andrew R. Conn	Nick Gould
IBM Watson Research Center	Rutherford Appleton Laboratory
Yorktown Heights, NY 10598-0218	Chilton, Oxfordshire
	England, United Kingdom
arconn@watson.ibm.com	nimg@ib.rl.ac.uk

References

A. R. Conn, N. I. M. Gould, and Ph. L. Toint, *A globally convergent augmented Lagrangian algorithm for optimization with general constraints and simple bounds*, SIAM J. Numer. Anal., 28 (1991), pp. 545–572.

———, LANCELOT: *A Fortran package for large-scale nonlinear optimization (Release A)*, Springer Series in Computational Mathematics 17, Springer-Verlag, New York, 1992.

LBFGS

Areas covered by the software
Large-scale unconstrained minimization

Basic algorithms
The method implemented in LBFGS is the limited-memory BFGS algorithm, as described by Liu and Nocedal. It is intended for problems with many variables. In this method quasi-Newton corrections are stored in vector form;

when the available storage is used up, the oldest correction is deleted to make space for a new one. The user specifies the number m of BFGS corrections that should be stored. LBFGS requires $2m(n + 1) + 4n$ storage locations.

The step length is determined at each iteration by a line-search routine (supplied by Moré and Thuente) that enforces a sufficient decrease condition and a curvature condition.

Hardware/software environment

LBFGS is written in Fortran 77. Single- and double-precision versions of the software are available. Machine dependencies are restricted to BLOCK DATA LB2.

Contact address

>Jorge Nocedal
>Northwestern University
>Department of Electric Engineering and Computer Science
>Evanston, IL 60208
>Phone: (708) 491-5038
>nocedal@eecs.nwu.edu

Additional comments

LBFGS is also available in the Harwell Library under the name VA15.

References

D. C. Liu and J. Nocedal, *On the limited memory BFGS method for large-scale optimization*, Math. Programming, 45 (1989), pp. 503–528.

J. Nocedal, *Updating quasi-Newton matrices with limited storage*, Math. Comp., 24 (1980), pp. 773–782.

LINDO

Areas covered by the software

Linear programming, mixed-integer linear programming, quadratic programming

Basic algorithms

LINDO was originally developed by Schrage. It uses simplex and active set algorithms for linear and quadratic programming, and a branch-and-bound approach for mixed-integer programming.

SOFTWARE PACKAGES

Hardware/software environment

LINDO is available in a number of formats that have different platform requirements. These are summarized in the table below.

	LINDO	Super LINDO	Hyper LINDO	Industrial LINDO	Extended LINDO
Variables	200	1,000	4,000	16,000	100,000
Constraints	100	500	2,000	8,000	32,000
Nonzeros	4,000	16,000	64,000	200,000	1,000,000
Memory^a	640,000	640,000	2MB	5MB	16MB
Memory^b	1MB	1MB	2MB	n.s.	n.s.
Memory^c	n.s.	n.s.	2MB	5MB	16MB

^{a}PC; ^{b}Macintosh; ^{c}Workstation; n.s. = not supported.

Mainframe versions are also available. The workstations supported are DEC 3100, DEC 5000, HP, IBM RS/6000, MicroVAX, NeXT, Sun SPARCstation, and Sun 3. Linkable Fortran object code is supplied with Hyper, Extended, and Industrial LINDO. MPS and Fortran interfaces are available for data entry.

LINDO supports standard operating system environments for the hardware described above. The PC version includes a full-screen editor and a pop-up window that allows the progress of the algorithm to be monitored.

Contact address

LINDO Systems, Inc.
P. O. Box 148231
Chicago, IL 60614
Phone: (800) 441-BEST and (312) 871-2524
Fax: (312) 871-1777

Reference

LINDO: An Optimization Modeling System, Text and Software, 4th ed., The Scientific Press, San Francisco, CA, 1988.

LINGO

Areas covered by the software

LINGO is a modeling language that allows linear and integer-programming models to be set up and solved efficiently. A library of mathematical, probability, and financial functions is included.

Basic algorithms

LINGO interfaces with the LINDO package.

Hardware/software environment

LINGO is available in a number of formats that have different platform requirements. These are summarized in the table below.

	LINGO	Super LINGO	Hyper LINGO	Industrial LINGO	Extended LINGO
Variables	200	1,000	4,000	16,000	100,000
Constraints	100	500	2,000	8,000	32,000
Nonzeros	4,000	16,000	64,000	200,000	800,000
Memory^a	640,000	640,000	2MB	5MB	16MB
Memory^b	1MB	1MB	2MB	n.s.	n.s.
Memory^c	n.s.	n.s.	2MB	5MB	16MB

^{a}PC; ^{b}Macintosh; ^{c}Workstation; n.s. = not supported.

Mainframe versions are also available. The workstations supported are DEC 3100, DEC 5000, HP, IBM RS/6000, MicroVAX, NeXT, Sun SPARCstation, Sun3. A full-screen editor is available for easy model generation. Model data can be stored in spreadsheet files. Interfaces for MPS files and LINDO TAKE files are also available.

Contact address

LINDO Systems, Inc.
P. O. Box 148231
Chicago, IL 60614
Phone: (800) 441-BEST and (312) 871-2524
Fax: (312) 871-1777

LNOS

Areas covered by the software

The classical linear programming/network flow problems: shortest path, maximum flow, assignment, minimum-cost flow

Basic algorithms

The LNOS (Linear Network Optimization Software) package contains the following codes:

GRIDGEN—Generates minimum-cost flow problems with a grid structure.

NGCONVERT—Converts problems from the NETGEN format to a standard format.

SOFTWARE PACKAGES

PAPE_ALL_DEST—Pape's method for shortest paths from one origin to all destinations.

HEAP_ALL_DEST—Binary heap method for shortest paths from one origin to all destinations.

HEAP_SELECT_DEST—Binary heap method for shortest paths from one origin to selected destinations.

AUCTION_SELECT_DEST—Auction algorithm for shortest paths from one origin to selected destinations.

RELAX—Relaxation method for minimum-cost flow problems.

AUCTION—Forward auction algorithm for symmetric assignment problems.

AUCTION_FLP—Same as AUCTION but uses floating-point arithmetic to deal with problems with large cost range and/or dimension.

AUCTION_AS—Auction algorithm for asymmetric assignment problems.

AUCTION_FR—Forward/reverse auction algorithm for symmetric assignment problems.

NAUCTION_SP—Combined naive auction and sequential shortest-path method for assignment.

FORD_FULKERSON—Ford–Fulkerson algorithm for maximum flow.

E_RELAX_MF—E-relaxation algorithm for maximum flow.

E_RELAX—E-relaxation algorithm for minimum-cost flow.

E_RELAX_N—E-relaxation algorithm for minimum-cost flow; iterates from both positive and negative surplus nodes.

Hardware/software environment

All codes are written in Fortran and, with the exception of AUCTION_FLP, use integer arithmetic. All codes can be compiled with the popular Absoft compiler on all Macintosh computers. By changing a few input and output statements, the codes can be easily adapted for other computers and Fortran compilers.

Contact address

Listings of all codes except RELAX, AUCTION_FLP, and E_RELAX_N appear in the book cited below. For an IBM PC- or Macintosh-readable diskette with the listings of all the codes, send $25.00 to:

>Dimitri P. Bertsekas
>Massachusetts Institute of Technology
>Room 35-210
>Cambridge, MA 02139

Reference

D. P. Bertsekas, *Linear Network Optimization: Algorithms and Codes*, MIT Press, Cambridge, MA, 1991.

LPsolver

Areas covered by the software

Linear programming

Basic algorithms

LPsolver provides two algorithms for solving a linear program: *Stepwise Solve* and *Revised Simplex Solve*. Stepwise Solve is primarily intended for instructional purposes and acts much like a linear programming calculator. It prompts the user for the entering and leaving basic variables, and it performs the necessary calculations based on the user's selection. The Revised Simplex Solve uses the revised simplex algorithm to solve the problem automatically.

Hardware/software environment

The program is written in Microsoft QuickBasic 4.5. The software runs on any IBM-compatible computer.

Contact addresses

> Paul A. Savory
> Department of Industrial Engineering
> Arizona State University
> Tempe, AZ 85287-5906
> Phone: (602) 965-3185
> aspfs@asuvm.inre.asu.edu

> Jeffrey L. Arthur
> Department of Statistics
> Oregon State University
> Corvallis, OR 97331-4606
> Phone: (503) 737-2429
> arthur@stat.orst.edu

References

K. Murty, *Linear and Combinatorial Programming*, John Wiley & Sons, Inc., New York, 1976.

P. Savory and J. Arthur, *LPSOLVER—A Linear Program Solving Package for Educational and Instructional Modeling Applications*, Tech. Report 142, Department of Statistics, Oregon State University, Corvallis, OR, 1990.

LSGRG2

Areas covered by the software

Nonlinear programming

SOFTWARE PACKAGES

Basic algorithms

LSGRG2 uses an implementation of the generalized reduced-gradient (GRG) algorithm similar to that used in GRG2. However, it uses a sparse data structure to store and manipulate the constraint Jacobian matrix and uses a sparse inversion procedure to factor the basis. It can therefore solve large, sparse nonlinear programs: problems with over 500 constraints have been solved successfully.

LSGRG2 requires the same user-supplied subroutines as GRG2 and has similar subroutine and data file interfaces. See the GRG2 entry for details.

Hardware/software environment

LSGRG2 is written in ANSI Fortran. Machine dependencies are relegated to the subroutine INITLZ, which defines three machine-dependent constants.

Contact address

> Leon Lasdon
> MSIS Department
> College of Business Administration
> University of Texas
> Austin, TX 78712-1175
> Phone: (512) 471-9433

Additional comments

LSGRG2 has been interfaced with the GAMS modeling language. For tests comparing LSGRG2 with MINOS on a set of large GAMS models, see the reference below.

Reference

S. Smith and L. Lasdon, *Solving large sparse nonlinear programs using GRG*, ORSA J. Comput., 4 (1992), pp. 1–15.

LSNNO

Areas covered by the software

Optimization problems with nonlinear objective, bound constraints, and linear network constraints, with an emphasis on large-scale problems

Basic algorithms

LSNNO (Large Scale Nonlinear Network Optimizer) is a line-search method where the search direction is obtained by a truncated conjugate gradient technique. The bounds are handled by bending the search direction on the boundary of the feasible domain. It features both Newton and quasi-Newton algorithms.

LSNNO exploits the partially separable structure of many large-scale problems to obtain good efficiency on such problems. In particular, it uses the partitioned updating technique when a quasi-Newton method is chosen.

Hardware/software environment

LSNNO is a standard ANSI Fortran subroutine in double precision. It uses a reverse communication interface for the calculation of objective function values and constraint values and their derivatives.

Contact address

>Philippe Toint
>Department of Mathematics
>FUNDP
>61, rue de Bruxelles
>B5000 Namur, Belgium
>pht@math.fundp.ac.be

Additional comments

The package was written by Tuyttens.

References

Ph. L. Toint and D. Tuyttens, *On large-scale nonlinear network optimization*, Math. Programming B, 48 (1990), pp. 125–159.

———, *LSNNO: A Fortran subroutine for solving large-scale nonlinear network optimization problems*, ACM Trans. Math. Software, 18 (1992), pp. 308–328.

LSSOL

Areas covered by the software

LSSOL is a Fortran package for linearly constrained linear least squares problems and convex quadratic programming. LSSOL is designed to solve a class of linear and quadratic programming problems of the following general form:

$$\minimize_{x \in \mathbb{R}^n} \quad f(x)$$
$$\text{subject to} \quad \ell \leq \left\{ \begin{array}{c} x \\ Cx \end{array} \right\} \leq u,$$

where C is an $m_L \times n$ matrix (m_L may be zero) and $f(x)$ is one of the following:

FP: *None*   (find a feasible point for the constraints)

LP: $c^T x$ (a linear program)

QP1: $\frac{1}{2} x^T A x$ A symmetric and positive semidefinite,

QP2: $c^T x + \frac{1}{2} x^T A x$ A symmetric and positive semidefinite,

QP3: $\frac{1}{2} x^T A^T A x$ A $m \times n$ upper trapezoidal,

QP4: $c^T x + \frac{1}{2} x^T A^T A x$ A $m \times n$ upper trapezoidal,

LS1: $\frac{1}{2} \|b - Ax\|^2$ A $m \times n$,

LS2: $c^T x + \frac{1}{2} \|b - Ax\|^2$ A $m \times n$,

LS3: $\frac{1}{2} \|b - Ax\|^2$ A $m \times n$ upper trapezoidal,

LS4: $c^T x + \frac{1}{2} \|b - Ax\|^2$ A $m \times n$ upper trapezoidal,

with c an n-vector and b an m-vector.

Basic algorithms

LSSOL uses a two-phase, active-set method related to the method used in the package QPOPT. Two special features of LSSOL are its exploitation of convexity and treatment of singularity. LSSOL treats all matrices as dense and is not intended for sparse problems.

Hardware/software environment

The LSSOL package contains approximately 15,000 lines of Fortran, of which about 75% are comments. The source code and example program for LSSOL are distributed on a floppy disk. A MATLAB interface for LSSOL is also available.

LSSOL includes calls to both Level-1 (vector) and Level-2 (matrix-vector) Basic Linear Algebra Subroutines (BLAS). They may be replaced by versions of the BLAS routines that have been tuned to a particular machine.

LSSOL is written in ANSI Fortran 77 and should run without alteration on any machine with a Fortran 77 compiler. The code was developed on a DECstation 3100 using the MIPS f77 compiler and has been installed on most types of PC, workstation, and mainframe.

Contact addresses

          Stanford Business Software, Inc.
          2672 Bayshore Parkway, Suite 304
          Mountain View, CA 94043
          Phone: (415) 962-8719
          Fax: (415) 962-1869

Additional comments

LSSOL is essentially identical to the routine E04NCF of the NAG Fortran Library. E04NCF was introduced at Mark 13. LSSOL was first distributed by the Office of Technology Licensing at Stanford in 1986. Since that time, the routine has been continually revised. Users with older versions of LSSOL should consider obtaining a copy of the most recent version.

References

P. E. Gill, S. J. Hammarling, W. Murray, M. A. Saunders, and M. H. Wright, *User's Guide for LSSOL (Version 1.0): A Fortran Package for Constrained Linear Least-Squares and Convex Quadratic Programming*, Tech. Report SOL 86-1, Department of Operations Research, Stanford University, Stanford, CA, 1986.

The NAG Fortran Library, Numerical Algorithms Group Limited, Wilkinson House, Jordan Hill Road, Oxford, England, UK, 1993.

J. Stoer, *On the numerical solution of constrained least-squares problems*, SIAM J. Numer. Anal., 8 (1971), pp. 382–411.

M1QN2 and M1QN3

Areas covered by the software

Large-scale unconstrained optimization

Basic algorithms

The routines are based on limited memory quasi-Newton methods (the LBFGS method of Nocedal) with a dynamically updated scalar (M1QN2) or diagonal (M1QN3) preconditioner. The Wolfe line search is used, and the step size is determined by the Fletcher–Lemaréchal algorithm.

Hardware/software environment

The software is written in Fortran 77. Single- and double-precision versions are available.

Contact addresses

Both Claude Lemaréchal and Jean Charles Gilbert can be reached at the following address:

> INRIA
> Domaine de Voluceau, BP 105
> 78153 Rocquencourt
> France

SOFTWARE PACKAGES

>Phone (Lemaréchal): +33 (1) 39.63.56.81
>`lemarech@seti.inria.fr`
>Phone (Gilbert): +33 (1) 39.63.55.24
>`gilbert@seti.inria.fr`

Additional comments

The routines are part of the MODULOPT library and use its communication protocol. They are fully documented.

Reference

J. Ch. Gilbert and C. Lemaréchal, *Some numerical experiments with variable-storage quasi-Newton algorithms*, Math. Programming, 45 (1989), pp. 407–435.

MATLAB Optimization Toolbox

Areas covered by the software

Linear programming, quadratic programming, unconstrained and constrained optimization of nonlinear functions, nonlinear equations, nonlinear least squares, minimax, multiobjective optimization, and semi-infinite programming

Basic algorithms

Linear programming—A variant of the simplex method. An initial phase is needed to identify a feasible point.

Quadratic programming—An active set method. A linear programming problem is solved to determine an initial feasible point.

Unconstrained minimization—Two routines are supplied. One implements a quasi-Newton algorithm, using either DFP or BFGS to update an approximate inverse Hessian, according to a switch selected by the user. Gradients may be supplied by the user; if they are not, finite differencing is used. The second routine uses the Nelder–Mead simplex algorithm, for which derivatives are not needed.

Constrained minimization—Sequential quadratic programming. The BFGS formula is used to maintain an approximation to the Hessian. Han's merit function is used to determine the step length at each iteration.

Nonlinear equations—Newton's method and the Levenberg–Marquardt algorithm are supplied. The user chooses the algorithm by setting a switch.

Nonlinear least squares—The Gauss–Newton method and the Levenberg–Marquardt method are supplied. The user makes the choice.

Minimax—These problems can be formulated as constrained optimization problems, and a sequential quadratic programming algorithm is used to solve them here. Advantage is taken of the structure of the problem in the choice of approximate Hessian.

Multiobjective optimization—The problem is formulated as one of decreasing a number of objective functions below a certain threshold simultaneously, so it is viewed as a constrained optimization problem. Again, sequential quadratic programming is used to solve it.

Semi-infinite programming—Cubic and quadratic interpolation is used to locate peaks in the infinite constraint set and therefore to reduce the problem to a constrained optimization problem.

Hardware/software environment

The software is made available as MATLAB M-files. It executes on any environment that supports MATLAB (including most PC and workstation environments in common use, as well as the Alliant family and the Cray UNICOS environment). The user supplies the function and constraint information to the toolbox as an M-file or as a series of MATLAB statements. If the functions are too complicated to be expressed as MATLAB statements, the MATLAB facilities for interfacing to Fortran and C routines can be invoked.

Contact address

> The MathWorks, Inc.
> Cochituate Place
> 24 Prime Park Way
> Natick, MA 01760
> Phone: (508) 653-1415
> Fax: (508) 653-2997
> info@mathworks.com

Reference

A. Grace, *Optimization Toolbox User's Guide*, The MathWorks, Inc., Natick, MA, 1990.

MINOS

Areas covered by the software

Linear programming, unconstrained and constrained nonlinear optimization

Basic algorithms

Linear programming—A primal simplex method is used. A sparse *LU* factorization of the basis matrix is maintained, using the Markowitz ordering scheme and Bartels–Golub updates as implemented in the LUSOL package of Gill, Murray, Saunders, and Wright.

Nonlinear objective, linear constraints—A reduced-gradient algorithm is used in conjunction with a quasi-Newton algorithm. The constraints are handled by an active set strategy. Feasibility is maintained throughout the process. The variables are classified as *basic, superbasic,* and *nonbasic*; at the solution, the basic and superbasic variables are away from their bounds. The null space is spanned by a matrix that is constructed from the coefficient matrix of the basic variables by using a sparse factorization. On the active constraint manifold, a quasi-Newton approximation to the reduced Hessian is maintained.

Nonlinear programming—A projected augmented Lagrangian algorithm is used. Feasibility is maintained with respect to the linear constraints. At each iteration, a linearly constrained subproblem is solved: the nonlinear constraints are linearized, and the objective function is augmented by a Lagrangian term and a quadratic penalty term for the nonlinear constraints.

Hardware/software environment

MINOS 5.4 is written in standard Fortran 77.

Contact address

Stanford Business Software, Inc.
2672 Bayshore Parkway, Suite 304
Mountain View, CA 94043
Phone: (415) 962-8719
Fax: (415) 962-1869

Additional comments

Input to MINOS 5.4 is via MPS files (which contain information for the linear parts of the objective function and constraints), SPECS files (which specify the problem types and set various parameters), and Fortran codes (which calculate objective and constraint functions, if nonlinear). The GAMS system can be used as an alternative user interface. See the entry on GAMS for details.

MINOS was developed by Murtagh and Saunders.

Reference

B. A. Murtagh and M. A. Saunders, *MINOS 5.1 User's Guide*, Tech. Report SOL 83–20R, Systems Optimization Laboratory, Stanford University, Stanford, CA, December 1983; revised January 1987.

MINPACK-1

Areas covered by the software
Systems of nonlinear equations and nonlinear least squares problems

Basic algorithms
The algorithms in MINPACK-1 are based on the trust-region concept. A modification of Powell's hybrid algorithm is used for systems of nonlinear equations, and a version of the Levenberg–Marquardt algorithm is used for nonlinear least squares problems.

For each problem area there are algorithms that proceed from the analytic Jacobian matrix or directly from the functions themselves. Since the specification of the Jacobian matrix can be an error-prone task, MINPACK-1 also contains an algorithm to check that the Jacobian matrix is consistent with the functions. Also included in the package are machine-readable documentation and a complete set of testing aids.

Hardware/software environment
Software is written in ANSI Fortran. Single- and double-precision versions of the software are available. Machine dependencies are restricted to a single subroutine that defines three machine-dependent constants.

Contact addresses
Software can be obtained from *netlib* (send index from minpack). Other sources include

C. Abaci, Inc.	NAG, Inc.
208 St. Mary's St.	1400 Opus Place
Raleigh, NC 27605	Downers Grove, IL 60515
Phone: (919) 832-4847	Phone: (708) 971-2337

Additional comments
This package is currently being revised and expanded.

References

J. J. Moré, B. S. Garbow, and K. E. Hillstrom, *User Guide for MINPACK-1*, Argonne National Laboratory Report ANL-80-74, Argonne, IL, 1980.

J. J. Moré, D. C. Sorensen, K. E. Hillstrom, and B. S. Garbow, *The MINPACK Project*, in Sources and Development of Mathematical Software, W. J. Cowell, ed., Prentice–Hall, Englewood Cliffs, NJ, 1984.

SOFTWARE PACKAGES

MIPIII

Areas covered by the software
Mixed-integer programming models

Basic algorithms
MIPIII is an implementation of the branch-and-bound algorithm, using C-WHIZ to solve the continuous relaxation at each node of this procedure. MIPIII accepts six different kinds of discrete variables: binary, general integer, bivalent, semicontinuous, and special ordered sets of type 1 and type 2. MIPIII incorporates extensive heuristics to aid in quickly identifying "good" integer solutions, and provides an extensive list of user-settable parameters that can aid in solution. MIPIII accepts MPS standard input files with integer markers or with integer variable lists.

Hardware/software environment
MIPIII is written in DATAFORM and can currently be run on PCs under Extended DOS, the IBM RS/6000, Sun, and HP 700 series UNIX workstations, and IBM and IBM-compatible mainframes.

Contact addresses

Ketron Management Science
1700 N. Moore St.
Suite 1710
Arlington, VA 22209
Phone: (703) 558-8701

R. Staudinger
ARBED Information Systems
19, avenue de la Liberté
Luxembourg
Phone: +352 4792-2107

Additional comments
MIPIII is undergoing continual enhancement, and an ANSI C version is being prepared.

Reference
MIPIII User's Manual, Ketron Management Science, Arlington, VA, 1989.

MODULOPT

Areas covered by the software
Unconstrained problems and problems with simple bounds

Contents
This collection consists of two subcollections of Fortran programs: test problems and algorithms. It is filled and maintained by members of a French

club (the modulopt club), which includes both users of optimization and algorithm designers. As a result, the test problems are usually "real-life" ones, possibly experimental. There is a well-defined format for communication between test problems and algorithms. The standard is at present based on Fortran 77, and a major requirement of the programs is that they be highly portable.

Contact addresses

Both Claude Lemaréchal and Jean Charles Gilbert can be reached at the following address:

>INRIA
>Domaine de Voluceau, BP 105
>78153 Rocquencourt
>France
>Phone (Lemaréchal): +33 (1) 39.63.56.81
>lemarech@seti.inria.fr
>Phone (Gilbert): +33 (1) 39.63.55.24
>gilbert@seti.inria.fr

NAG C Library

Areas covered by the software

Chapter E04 of this library covers linear programming, quadratic programming, minimization of a nonlinear function (unconstrained or bound-constrained), minimization of a sum of squares.

Basic algorithms

For unconstrained problems and problems with simple bounds, quasi-Newton and conjugate gradient methods are provided. For minimizing a sum of squares, a Gauss–Newton method is used. The LP and QP routines use a numerically stable active set strategy.

An option-setting mechanism is provided in all routines, in order to keep the basic parameter list to a minimum, while allowing a large degree of flexibility in controlling the algorithm. The routines have the ability to print the solution, as well as various amounts of intermediate output to monitor the computation.

Service routines are provided for checking user-supplied routines for first derivatives and for computing a covariance matrix for nonlinear least squares problems.

Hardware/software environment

The NAG C Library is available in tested, compiled form for several hardware/software computing environments.

SOFTWARE PACKAGES

Contact addresses

NAG, Inc.
1400 Opus Place
Suite 200
Downers Grove, IL 60515-5702
Phone: (708) 971-2337
Fax: (708) 971-2706

NAG Ltd.
Wilkinson House, Jordan Hill Road
Oxford OX2 8DR
England, United Kingdom
Phone: +44 (865) 511245
Fax: +44 (865) 310139

NAG GmbH
Schleissheimerstrasse 5
W-8046 Garching bei München
Germany
Phone: +49 (89) 3207395
Fax: +49 (89) 3207396

Reference

NAG C Library Manual, Mark 2, NAG Ltd., Oxford, England, UK, 1992.

NAG Fortran Library

Areas covered by the software

Chapters E04 and H cover linear programming; mixed-integer linear programming; quadratic programming; minimization of a nonlinear function (unconstrained, bound constrained, linearly constrained, and nonlinearly constrained); and minimization of a sum of squares (unconstrained, bound constrained, linearly constrained, and nonlinearly constrained).

Basic algorithms

For problems with nonlinear constraints, a sequential QP algorithm is used. For unconstrained problems and problems with simple bounds, quasi-Newton, modified Newton, and conjugate gradient methods are provided. The Nelder–Mead simplex method is provided for unconstrained problems. For minimizing a sum of squares, a Gauss–Newton method is used. The LP and QP routines use a numerically stable active set strategy.

An option-setting mechanism is provided in the more recent routines, in order to keep the basic parameter list to a minimum, while allowing a large degree of flexibility in controlling the algorithm. These routines also have the ability to print the solution, as well as various amounts of intermediate output to monitor the computation.

Service routines are provided for approximating first or second derivatives by finite differences, for checking user-supplied routines for first or second derivatives, and for computing a covariance matrix for nonlinear least squares problems.

Hardware/software environment

The NAG Fortran Library is available in tested, compiled form for a large number of different hardware and software computing environments.

Contact addresses

NAG, Inc.
1400 Opus Place
Suite 200
Downers Grove, IL 60515-5702
Phone: (708) 971-2337
Fax: (708) 971-2706

NAG Ltd.
Wilkinson House, Jordan Hill Road
Oxford OX2 8DR
England, United Kingdom
Phone: +44 (865) 511245
Fax: +44 (865) 310139

NAG GmbH
Schleissheimerstrasse 5
W-8046 Garching bei München
Germany
Phone: +49 (89) 3207395
Fax: +49 (89) 3207396

Additional comments

The NAG Fortran Library is updated to a new Mark from time to time. Mark 16, which contains improved routines for linear and quadratic programming was released in the first half of 1993. Further developments in optimization routines are planned for future Mark versions.

Reference

NAG Fortran Library Manual, Mark 16, NAG Ltd., Oxford, England, UK, 1993.

NETFLOW

Areas covered by the software

Network optimization problems, including minimum-cost flow, maximum flow, and matching problems. Also, test problems arising from applications and codes for generating random network problems.

Basic algorithms

A wide variety of algorithms is implemented. For minimum-cost flow, the simplex algorithm of Kennington and Helgason is available. For maximum flow problems, there are solvers due to Goldberg, Waissi, and Dinic. Algorithms due to Gabow and Micali and Vazirani are available for matching problems.

Users are advised to browse through the directory (and, in particular, the file `directory.notes` and the `readme` files on each subdirectory) for information on algorithms, authors, references, languages, and input formats.

Hardware/software environment
Codes in Fortran, C, and Pascal are present. Most codes utilize the DIMACS standard input/output format.

Contact address
This collection of public-domain software is available via anonymous ftp from DIMACS, the Center for Discrete Mathematics and Theoretical Computer Science. Users with access to a Unix machine with an Internet connection can ftp to the address `dimacs.rutgers.edu`, login as `anonymous`, and supply their electronic mail address as a password. The subdirectory `pub/netflow` contains the codes discussed above. The contact address of DIMACS is

> DIMACS Center
> Rutgers, the State University of New Jersey
> P.O. Box 1179
> Piscataway, NJ 08855-1179
> Phone: (908) 932-5928
> Fax: (908) 932-5932
> center@dimacs.rutgers.edu

Additional comments
This software is in the public domain. It is not maintained by DIMACS and comes with no guarantees.

References
A number of authors are represented by the codes in this collection. Documentation in the anonympus ftp area contains specific references.

NETSOLVE

Areas covered by the software
Interactive package for linear network optimization problems

Basic algorithms
The package uses standard algorithms to solve assignment, maximum-flow, minimum-cost flow, longest- and shortest-path, minimum spanning tree, transportation, traveling salesman, and matching problems. Network topology is input interactively by the user, using descriptive names for network nodes and edges. Problems are solved by entering the problem name (for example, MINFLOW to solve the minimum-cost flow problem).

Hardware/software environment
For use with IBM and IBM-compatible PCs running DOS 2.1.

Contact address
>Upstate Resources, Inc.
>P.O. Box 152
>Six Mile, SC 29682

Additional comments
The package is issued in conjunction with the book referenced below. The standard version can accommodate problems with up to 50 nodes and 200 edges. An extended version, which handles up to 350 nodes and 1000 edges, can be obtained by writing to the contact address.

Reference
E. Minieka, *Optimization Algorithms for Networks and Graphs*, Marcel Dekker, New York, 1978.

NITSOL (version 1)

Areas covered by the software
Large-scale systems of nonlinear equations

Basic algorithms
NITSOL implements a globalized Newton iterative method to determine an approximate zero of a given function. Restarted GMRES is used to obtain approximate solutions of the linear systems that characterize Newton steps; GMRES iterations are terminated when the linear residual norm satisfies an inexact Newton condition. Globalization is by safeguarded backtracking as outlined by Eisenstat and Walker.

Hardware/software environment
The present code is in double-precision Fortran.

Contact address
>Homer Walker
>Mathematics and Statistics Department
>Utah State University
>Logan, UT 84322-3900
>walker@math.usu.edu

Additional comments

NITSOL is at the first stage of development. Future versions may incorporate alternative iterative linear solvers, alternative globalizations, and other features.

References

S. C. Eisenstat and H. F. Walker, *Globally Convergent Inexact Newton Methods*, Res. Report February/91/51, Mathematics and Statistics Department, Utah State University, Logan, UT, 1991.

H. F. Walker, *NITSOL (version 1): A GMRES-backtracking Newton iterative solver*, Res. Report (to appear), Mathematics and Statistics Department, Utah State University, Logan, UT, 1993.

NLPE

Areas covered by the software

The package is described in the book *Nonlinear Parameter Estimation: An Integrated System in BASIC*, cited below. The source code on the diskette accompanying this book is designed to solve function minimization and nonlinear least squares problems from a wide variety of fields. An extensions diskette includes a large number of example problem files that can be modified for use by the reader.

Basic algorithms

Minimization algorithms—The original diskette includes the following methods: Hooke and Jeeves, Nelder–Mead, conjugate gradient, variable metric, truncated Newton, and a modified Marquardt nonlinear least squares method. The extensions diskette adds a safeguarded Newton method (using second-derivative information). All codes accept upper and lower bounds on the parameters to be determined.

Support codes—These include a driver, post-solution analysis (for parameter sensitivity and variance estimation), plotting routines, and environment determination. Many problem files are included.

Program builders—The original disk includes the BATch file NL.BAT. By typing NL (method_name) (problem_name) , the user causes a program to solve the named problem by the given method. Sensible defaults are provided for most cases.

Hardware/software environment

The software is intended to run on IBM PC and compatible computers, but since its source code is written in a close approximation to ISO Minimal BASIC, it has been run in parts on other platforms. Users are generally expected to modify a problem file to adapt existing examples to their needs.

Contact address

The source code diskette is included with the book cited below. A freeware subset of the codes is available as a UUencoded self-extracting PC file, which can be obtained by email. The extensions diskette can be obtained from

>Nash Information Services Inc.
>1975 Bel Air Drive
>Ottawa, Ontario, K2C 0X1
>Canada
>jxnhg@acadvm1.uottawa.ca

Reference

J. C. Nash and M. Walker-Smith, *Nonlinear Parameter Estimation: An Integrated System in BASIC*, Marcel Dekker, New York, 1987.

NLPQL

Areas covered by the software

NLPQL solves general nonlinear mathematical programming problems with equality and inequality constraints

Basic algorithms

The internal algorithm is a sequential quadratic programming method. Proceeding from a quadratic approximation of the Lagrangian function and a linearization of the constraints, a quadratic subproblem is formulated and solved to get a search direction. Subsequently a line search is performed with respect to two alternative merit functions, and the Hessian approximation is updated by the modified BFGS formula.

Hardware/software environment

NLPQL is written in double-precision Fortran 77 and organized in the form of a subroutine. Nonlinear problem functions and analytical gradients must be provided by the user within special subroutines or the calling program.

Contact address

>Klaus Schittkowski
>Mathematisches Institut
>Universität Bayreuth
>8580 Bayreuth, Germany
>Klaus.Schittkowski@uni-bayreuth.de

Additional comments

The special features of NLPQL are separate handling of upper and lower bounds on the variables, reverse communication, internal scaling, initial multiplier and Hessian approximation, feasibility with respect to bounds and linear constraints, and full documentation by initial comments.

Reference

K. Schittkowski, *NLPQL: A Fortran subroutine for solving constrained nonlinear programming problems*, Ann. Oper. Res., 5 (1985/86), pp. 485–500.

NLPQLB

Areas covered by the software

NLPQLB extends the capabilities of the general nonlinear programming code NLPQL to problems with many constraints, where the derivative matrix of the constraints does not possess any special sparsity structure that can be exploited numerically.

Basic algorithms

The user defines the maximum number of lines in the matrix of the linearized constraints that can be stored in core. By investigating the constraint function values, a decision is made about which restrictions are necessary to fill that matrix. The algorithm will stop if too many constraints are violated.

Hardware/software environment

NLPQLB is a double-precision Fortran 77 subroutine in which all parameters are passed through subroutine arguments. The program organization is similar to that of NLPQL.

Contact address

> Klaus Schittkowski
> Mathematisches Institut
> Universität Bayreuth
> 8580 Bayreuth, Germany
> Klaus.Schittkowski@uni-bayreuth.de

Additional comments

In addition to the same special features as NLPQL (scaling, reverse communication, and so on), NLPQLB executes NLPQL in reverse communication and provides full documentation by initial comments.

References

K. Schittkowski, *NLPQL: A Fortran subroutine for solving constrained nonlinear programming problems*, Ann. Oper. Res., 5 (1985/86), pp. 489–500.

———, *Solving Nonlinear Programming Problems with Very Many Constraints*, Report No. 294, Schwerpunktprogramm "Anwendungsbezogene Optimierung und Steuerung," Mathematisches Institut, Universität Bayreuth, Bayreuth, Germany, 1991.

NLSFIT

Areas covered by the software

NLSFIT solves parameter-estimation problems in which the parameters are either part of an explicit model function or are hidden in an ordinary differential equation system. Measurements may be given for an arbitrary number of the individual model functions or compartments, respectively. Model changes are permitted, and initial values or breakpoints can be optimization variables. The problem may possess upper and lower bounds on the variables and, in addition, any smooth equality or inequality constraints.

Basic algorithms

The algorithm creates a nonlinear least squares problem and solves it by DFNLP. Differential equations are solved by an interface to six available differential equation solvers. For stiff or ill-conditioned systems, a shooting technique can be applied. Gradients are evaluated numerically, where the special data structures from shooting are taken into account.

Hardware/software environment

NLSFIT is a double-precision Fortran 77 subroutine, and all parameters are passed through arguments. An additional main program takes over some organizational ballast and reads in all problem data. A user-provided subroutine is required to define initial values, constraints, micro parameters, and the right-hand side of a differential equation system or, alternatively, an explicit model function.

Contact address

Klaus Schittkowski
Mathematisches Institut
Universität Bayreuth
8580 Bayreuth, Germany
Klaus.Schittkowski@uni-bayreuth.de

SOFTWARE PACKAGES

Additional comments

The special features are its scaling options (individual, within measurement sets, user-provided scaling factors, automatic scaling) and its interfaces with the least squares codes NLSNIP of Lindström and DN2GB of Gay and Welsh, as well as with differential equation solvers such as the IMSL routines DGEAR and DVERK. Full documentation by initial comments is provided.

References

K. Schittkowski, *GAUSS: Interactive Modelling and Parameter Estimation*, Report No. 293, Schwerpunktprogramm "Anwendungsbezogene Optimierung und Steuerung," Mathematisches Institut, Universität Bayreuth, Bayreuth, Germany, 1991.

——, *NLSFIT: A Fortran Code for Parameter Estimation in Differential Equations or Explicit Model Functions*, User Documentation, Mathematisches Institut, Universität Bayreuth, Bayreuth, Germany, 1990.

NLSSOL

Areas covered by the software

NLSSOL is a Fortran package designed to solve the constrained nonlinear least squares problem: the minimization of a sum of squares of smooth nonlinear functions subject to a set of constraints on the variables. The problem is assumed to be stated in the following form:

$$\minimize_{x \in I\!R^n} \quad \tfrac{1}{2} \sum_{i=1}^{m} (y_i - f_i(x))^2$$

$$\text{subject to} \quad \ell \leq \left\{ \begin{array}{c} x \\ Ax \\ c(x) \end{array} \right\} \leq u,$$

where the $f_i(x)$ are nonlinear functions, the y_i are constant, A is an $m_L \times n$ matrix, and $c(x)$ is an m_N-vector of nonlinear constraints.

The user must supply an initial estimate of the solution of the problem, subroutines that evaluate $f(x)$, $c(x)$, and as many first partial derivatives as possible; unspecified derivatives are approximated by finite differences.

Basic algorithms

NLSSOL is a sequential quadratic programming (SQP) method incorporating an augmented Lagrangian merit function and a BFGS quasi-Newton approximation to the Hessian of the Lagrangian. Features of NLSSOL include the explicit use of the Jacobian of $f(x)$ and the ability to define an initial approximate Hessian suitable for least squares problems.

NLSSOL utilizes subroutines from the NPSOL and LSSOL packages, which are distributed together with NLSSOL.

Hardware/software environment

NLSSOL contains approximately 16,500 lines of Fortran, of which about 75% are comments. The source code and example program for NLSSOL are distributed on a floppy disk. The code is also available via Internet using ftp.

NLSSOL includes calls to both Level-1 (vector) and Level-2 (matrix-vector) Basic Linear Algebra Subroutines (BLAS). They may be replaced by versions of the BLAS routines that have been tuned to a particular machine.

NLSSOL is written in ANSI Fortran 77 and should run without alteration on any machine with a Fortran 77 compiler. The code was developed on a DECstation 3100 using the MIPS f77 compiler and has been installed on most types of PC, workstation, and mainframe.

Contact addresses

Walter Murray
Department of Operations Research
Terman Engineering Center
Stanford University
Stanford, CA 94305-4022

Philip E. Gill
Department of Mathematics
University of California, San Diego
9500 Gilman Drive
La Jolla, CA 92093-0112

Additional comments

NLSSOL is essentially identical to the routine E04UPF of the NAG Fortran Library. E04UPF was introduced at Mark 14.

References

P. E. Gill, W. Murray, M. A. Saunders, and M. H. Wright, *Some theoretical properties of an augmented Lagrangian merit function*, in Advances in Optimization and Parallel Computing, P. M. Pardalos, ed., North–Holland, Amsterdam, 1992, pp. 101–128.

———, *User's Guide for NPSOL (Version 4.0): A Fortran Package for Nonlinear Programming*, Tech. Report SOL 86-2, Department of Operations Research, Stanford University, Stanford, CA, 1986.

NAG Fortran Library Manual, Mark 16, NAG Ltd., Oxford, England, UK, 1993.

NLPSPR

Areas covered by the software

NLPSPR is a Fortran package designed to solve the nonlinear programming problem when the number of variables and constraints is large and the Jacobian and Hessian matrices are sparse. The algorithm can be used to locate a feasible point for a system of nonlinear equality and/or inequality constraints. The algorithm is well suited for applications derived from discretized differential equations and boundary value problems.

Basic algorithms

NLPSPR is a sequential quadratic programming algorithm that uses an augmented Lagrangian merit function and a sparse quadratic programming algorithm based on the Schur complement approach of Gill, Murray, Saunders, and Wright. Sparse linear systems are solved efficiently using a multifrontal algorithm that implements a modified Cholesky decomposition for symmetric indefinite systems. The user must supply the sparse Jacobian and Hessian matrices, although this information can be computed efficiently using sparse finite differences which are implemented in a utility package that is also available. The software incorporates a reverse communication structure and is especially well suited for applications derived from discretized optimal control problems.

Hardware/software environment

The software is written in Fortran 77 and uses portions of the BCSLIB mathematical subroutine library, including the BLAS and LINPACK.

Contact address

> John T. Betts or Paul D. Frank
> Boeing Computer Services
> P.O. Box 24346, MS 7L-21
> Seattle, WA 98124-0346

Additional comments

The software is undergoing continuous revision, and availability is subject to negotiations with one of the authors and the Boeing Company.

Reference

J. T. Betts and P. D. Frank, *A Sparse Nonlinear Optimization Algorithm*, Tech. Report AMS-TR-173, Applied Mathematics and Statistics Group, Boeing Computer Services, Seattle, WA, 1991.

NPSOL

Areas covered by the software

NPSOL is a Fortran package designed to solve the nonlinear programming problem: the minimization of a smooth nonlinear function subject to a set of constraints on the variables. The problem is assumed to be stated in the following form:

$$\minimize_{x \in \mathbb{R}^n} \quad f(x)$$

$$\text{subject to} \quad \ell \leq \left\{ \begin{array}{c} x \\ Ax \\ c(x) \end{array} \right\} \leq u,$$

where $f(x)$ is a smooth nonlinear function, A_L is an $m_L \times n$ matrix, and $c(x)$ is an m_N-vector of smooth nonlinear constraint functions.

The user must supply an initial estimate of the solution of the problem, subroutines that evaluate $f(x)$, $c(x)$, and as many first partial derivatives as possible. Unspecified derivatives are approximated by finite differences.

If the problem is large and sparse, alternative software such as MINOS should be used instead, since NPSOL treats all matrices as dense.

Basic algorithms

NPSOL is a sequential quadratic programming method incorporating an augmented Lagrangian merit function and a BFGS quasi-Newton approximation to the Hessian of the Lagrangian. If there are no nonlinear constraints, the gradients of the bound and linear constraints are never recomputed, and NPSOL will function as a specialized algorithm for linearly constrained optimization.

It can be arranged that the problem functions are evaluated only at points that are feasible with respect to the bounds and linear constraints.

NPSOL uses subroutines from the LSSOL-constrained linear least squares package, which is distributed together with NPSOL.

Hardware/software environment

NPSOL contains approximately 16,500 lines of Fortran, of which about 75% are comments. The source code and example program for NPSOL are distributed on a floppy disk. A MATLAB interface for NPSOL is also available.

NPSOL includes calls to both Level-1 (vector) and Level-2 (matrix-vector) Basic Linear Algebra Subroutines (BLAS). They may be replaced by versions of the BLAS routines that have been tuned to a particular machine.

NPSOL is written in ANSI Fortran 77 and should run without alteration on any machine with a Fortran 77 compiler. The code was developed on a DECstation 3100 using the MIPS f77 compiler and has been installed on most types of PC, workstation, and mainframe.

SOFTWARE PACKAGES

Contact addresses

Stanford Business Software, Inc.
2672 Bayshore Parkway, Suite 304
Mountain View, CA 94043
Phone: (415) 962-8719
Fax: (415) 962-1869

Additional comments

NPSOL is essentially identical to the routine E04UCF of the NAG Fortran Library. E04UCF was introduced at Mark 13 and is currently available at Mark 16.

NPSOL was first distributed by the Office of Technology Licensing at Stanford in 1986. Since that time the routine has been continually revised. Users with older versions of NPSOL should consider obtaining a copy of the most recent version.

References

P. E. Gill, W. Murray, M. A. Saunders, and M. H. Wright, *Some theoretical properties of an augmented Lagrangian merit function*, in Advances in Optimization and Parallel Computing, P. M. Pardalos, ed., North–Holland, Amsterdam, 1992, pp. 101–128.

———, *User's Guide for NPSOL (Version 4.0): A Fortran package for nonlinear programming*, Tech. Report SOL 86-2, Department of Operations Research, Stanford University, Stanford, CA, 1986.

NAG Fortran Library Manual, Mark 16, NAG Ltd., Oxford, England, UK, 1993.

OB1 (Optimization with Barriers-1)

Areas covered by the software

Linear programming

Basic algorithms

The main algorithm is the primal-dual interior-point method, with Mehrotra's predictor-corrector strategy. The primal and dual simplex methods are also included. Either simplex method can be called after the interior method has converged, or they can be used independently. The interior-point method uses a supernodal sparse Cholesky factorization algorithm. Liu's multiple-minimum-degree algorithm or an implementation of the minimum-local-fill algorithm is used to reduce the fill-in in the Cholesky factor. A presolve algorithm is used to reduce problem size before solution, and a postsolve to construct an optimal solution to the original problem.

Hardware/software environment

The software is written entirely in ANSI Fortran. Double precision is used exclusively. Machine dependencies are restricted to the timer routine.

Contact addresses

Irvin Lustig
CPLEX Optimization, Inc.
99 Braeburn Drive
Princeton, NJ 08540
Phone: (609) 497-0984
irv@dizzy.cplex.com

Roy Marsten	David Shanno
ISYE	RUTCOR
Georgia Tech	Rutgers University
Atlanta, GA 30332	New Brunswick, NJ 08903
Phone: (404) 894-3983	Phone: (908) 932-4858

Additional comments

The software has been in development since 1987. Linear programs with up to 40,000 constraints and 180,000 variables have been solved.

References

I. J. Lustig, R. R. Marsten, and D. F. Shanno, *Computational experience with a primal-dual interior-point method for linear programming*, Linear Algebra Appl., 152 (1991), pp. 191–222.

——, *Implementing Mehrotra's predictor-corrector interior-point method for linear programming*, SIAM J. Optim., 2 (1993), pp. 435–449.

User's Manual for the OB1 Linear Programming System, XMP Software, Incline Village, NV, 1991.

ODRPACK

Areas covered by the software

Nonlinear least squares (NLS) and orthogonal distance regression (ODR) problems, including those with implicit models and multiresponse data

Basic algorithms

The NLS and ODR algorithms in ODRPACK are based on a trust-region Levenberg–Marquardt method. The procedure uses scaling to accommodate automatically problems in which the estimated values have widely varying magnitudes. The structure of the ODR problem is exploited so that the computational cost per iteration is equal to that of the NLS problem. When the model is implicit, the solution is found using the classic quadratic penalty function method.

For both NLS and ODR, the algorithm can approximate the necessary Jacobian matrices using either forward or central finite differences when the user does not supply the code to compute them. If the user does supply this code, it can be verified by a derivative-checking procedure. The ODRPACK weighting facility allows the user to compensate for correlation between the responses within a given observation with ease; it also allows the user to compensate for unequal precision between the observations. The covariance matrix and the standard errors of the estimators are provided optionally.

Hardware/software environment

ODRPACK is a portable ANSI Fortran subroutine library, with no restrictions on problem size other than those imposed by the memory limits of the machine on which it is installed. Both single- and double-precision versions are available. Machine dependencies are restricted to three integer constants that must be specified within two subroutines.

Contact address

The ODRPACK code is in the public domain and can be obtained via *netlib* (`send index from odrpack`) or from

>   Janet E. Rogers
>   Applied and Computational Mathematics Division
>   National Institute of Standards and Technology
>   325 Broadway
>   Boulder, CO 80303-3328
>   Phone: (303) 497-5114
>   jrogers@bldr.nist.gov

References

P. T. Boggs, R. H. Byrd, J. E. Rogers, and R. B. Schnabel, *User's Reference Guide for ODRPACK Version 2.01 Software for Weighted Orthogonal Distance Regression*, National Institute of Standards and Technology NISTIR 4834, Boulder, CO, 1992.

P. T. Boggs, R. H. Byrd, and R. B. Schnabel, *A stable and efficient algorithm for nonlinear orthogonal distance regression*, SIAM J. Sci. Statist. Comput., 8 (1987), pp. 1052–1078.

P. T. Boggs, J. R. Donaldson, R. H. Byrd, and R. B. Schnabel, *ODRPACK software for weighted orthogonal distance regression*, ACM Trans. Math. Software, 15 (1989), pp. 348–364.

OPSYC (OPtimisation de SYstèmes Creux)

Areas covered by the software

Nonlinearly constrained large-scale optimization, assuming availability of second derivatives

Basic algorithms

The algorithms use a sequential quadratic programming algorithm, using a Hessian given by the user. A sophisticated line search that avoids the Maratos effect is used. A penalty on the displacement is added when necessary.

This software has been applied to optimization of large electrical networks.

Hardware/software environment

Software is written in Fortran 77. Only a double-precision version is available.

Contact address

> F. Bonnans
> INRIA
> Domaine de Voluceau, BP 105
> 78153 Rocquencourt
> France
> Fax: +33 (1) 39.63.53.30

Additional comments

Documentation is included. This package is currently being revised and expanded.

References

G. Blanchon, J. F. Bonnans, and J. C. Dodu, *Optimisation des réseaux électriques de grande taille*, Lecture Notes in Control and Inform. Sci., 144 (1990), pp. 423–431.

J. F. Bonnans, *Asymptotic admissibility of the unit stepsize in exact penalty methods*, SIAM J. Control Optim., 27 (1989), pp. 631–641.

OptiA

Areas covered by the software

Unconstrained and constrained optimization, quadratic programming, global optimization, nonsmooth optimization, minimax optimization, multicriteria optimization. The package is aimed at smaller problems; at most 20 unknowns and 20 constraints are allowed in the shareware version. Changes to this limitation can be negotiated with the authors.

Basic algorithms

Many of the standard algorithms that have appeared in the literature of the last 25 years are represented. For unconstrained optimization, the derivative-free methods of Nelder–Mead and Powell are available, in addition to various Newton, quasi-Newton, and nonlinear conjugate gradient implementations, for which first derivatives are either supplied by the user or approximated by finite differencing. For constrained optimization, methods based on sequential quadratic programming and augmented Lagrangian techniques are available, together with Powell's TOLMIN algorithm (for linearly constrained problems). For quadratic programming, Powell's version of the Goldfarb–Idnani algorithm and an alternative method of Schittkowski are available. Options for global optimization include algorithms based on cluster analysis and Monte Carlo techniques. For minimax optimization, an interface to the FSQP code is available, together with other algorithms based on the constrained optimization techniques mentioned above. The available techniques for (possibly constrained) nonsmooth optimization include the bundle methods of Lemaréchal.

Hardware/software environment

The package is designed to run on a PC environment (286 and higher) under DOS (3.1 and higher). A math coprocessor is highly recommended. The user can supply a main program and subroutines to evaluate functions and gradients, and to monitor the execution of the algorithm. These subroutines can be written in standard Fortran 77, or in C or Pascal versions that work with Microsoft compilers. A menu-driven interface is available to provide templates for the user-supplied routines to monitor algorithm performance and assist in the choice of appropriate solution techniques. Use of this interface is demonstrated in the second reference below.

Contact address

Suggestions for further development of the system can be directed to

> J. Fidler
> Institute of Information Theory and Automation
> Academy of Sciences of the Czech Republic
> Pod vodárenskou věží 4
> 182 08 Prague
> Czech Republic
> Phone: +42 (2) 815-2222
> Fax: +42 (2) 847-452
> `fidler@cspgas11.bitnet`

Additional comments

The OptiA system is distributed on floppy disks (about 3 MB) in the form of executable modules and object code (methods). The completely revised UNIX version of OptiA system is now available and can be run on HP Apollo

9000/720 workstations under X-Windows. This version is easily adaptable to run on other machines (Sun, VAX, DEC).

References

J. Doležal and J. Fidler, *Dialogue System OptiA for Minimization of Functions of Several Variables. Version 3.0—Examples*, Institute of Information Theory and Automation, Czechoslovak Academy of Sciences, Prague, Czechoslovakia, July 1992.

———, *Dialogue System OptiA for Minimization of Functions of Several Variables. Version 3.0—User's Guide*, Institute of Information Theory and Automation, Czechoslovak Academy of Sciences, Prague, Czechoslovakia, July 1992.

OPTIMA Library

Areas covered by the software

Unconstrained optimization, constrained optimization, sensitivity analysis

Basic algorithms

OPVM—Unconstrained optimization, unstructured objective function; suitable for small problems.

OPVMB—Optimization subject to simple bounds.

OPLS—Unconstrained nonlinear least squares.

OPNL—Nonlinear equations; solved by minimizing the sum of squares of the residuals.

OPCG—Nonlinear conjugate gradient method.

OPODEU—Unconstrained optimization problems by tracing the solution curve of a system of ODEs (homotopy method).

OPTNHP—Unconstrained optimization using the truncated Newton method; no Hessian storage or calculation required.

OPRQP—Sequential quadratic programming, but superseded by the OPXRQP routine.

OPXRQP—A more efficient implementation of sequential quadratic programming; uses the EQP variant.

OPSQP—Another implementation of sequential quadratic programming, but uses the IQP variant, which gives rise to inequality-constrained subproblems.

SOFTWARE PACKAGES

OPALQP—Similar to OPSQP, but uses an augmented Lagrangian line-search function.

OPSMT—Nonlinear programming, using a SUMT technique.

OPIPF—Sequential minimization of a sequence of augmented Lagrangians.

OPODEC—Homotopy method: traces the solution curve of a system of ODEs.

OPSEN—Tests the sensitivity of the objective function around the optimal point of an unconstrained problem.

OPSEC—Like OPSEN, but for the solution of a constrained problem.

Hardware/software environment
Software is written in Fortran 77.

Contact address
>M. C. Bartholomew-Biggs
>Numerical Optimisation Center
>Hatfield, Hertfordshire AL10 9AB
>England, United Kingdom

Reference
OPTIMA Manual, Numerical Optimisation Center, University of Hertfordshire, England, UK, July 1989.

OPTPACK

Areas covered by the software
Unconstrained optimization and nonlinear constrained optimization with special software to handle bound constraints, linear equality constraints, and general nonlinear constraints

Basic algorithms
Unconstrained optimization is performed using the conjugate gradient algorithm. Constrained optimization is performed using a new scheme that combines multiplier methods with preconditioning and linearization techniques to accelerate convergence.

Hardware/software environment
The software is written in double-precision Fortran.

Contact address
The nonlinear conjugate gradient software is available from *netlib* (**send cg from napack**). The constrained optimization software can be obtained from

William W. Hager
Department of Mathematics
University of Florida
Gainesville, FL 32611
Phone: (904) 392-0286
hager@math.ufl.edu

Additional comments

The code is documented by internal comments. Research reports providing the theoretical basis for the algorithms are available on request. User feedback is much appreciated.

Reference

W. W. Hager, *Analysis and Implementation of a Dual Algorithm for Constrained Optimization*, Res. Report, Department of Mathematics, University of Florida, Gainesville, FL, 1990; J. Optim. Theory Appl., 79 (1993), to appear.

OSL

Areas covered by the software

Linear programming, convex quadratic programming, and mixed-integer programming problems

Basic algorithms

For linear programming, primal and dual versions of the simplex method are implemented. The user is allowed to adjust the pricing strategy. In addition, three interior-point solvers (one primal, one primal-dual path-following, and one primal-dual predictor-corrector) are available. For network linear programming, a specialized simplex method is used.

The quadratic programming solver uses a two-phase approach. In the first phase, successive linear programming approximations to the actual problem are used to generate approximate solutions. When these solutions are sufficiently close, the second phase—a standard active-set technique—is activated.

A branch-and-bound technique is used for mixed-integer programming. A simplex method is used to solve the linear programming subproblems.

Hardware/software environment

The Optimization Subroutine Library (OSL) is available for a variety of hardware environments, from PCs to mainframes, including IBM, Hewlett-Packard, Silicon Graphics, and Sun workstations.

Input can be passed to OSL routines through array arguments in the subroutine calling sequence. OSL also has input-output service routines that read data from MPS and Lotus 1-2-3 files. An OSL application program may be written in C, Fortan, PL/I, or APL2.

A graphical user interface (GUI) is available for the IBM RISC System/6000 workstation version of OSL. This interface allows the user to solve problems in a point-and-click environment. GUI generates application code in Fortran or C as part of its functionality.

Contact address

> OSL Development
> IBM Corporation, 85BA MS 276
> Kingston, NY 12401-1099
> Phone: (914) 385-5027
> Fax: (914) 385-4372
> osl@vnet.ibm.com

Additional comments

Volume 31 (1992) of the *IBM Systems Journal* contains eight articles related to OSL. An overview of the product may be found in the article cited below.

Reference

D. G. Wilson, *A Brief Introduction to the IBM Optimization Subroutine Library*, SIAG/OPT Views and News, 1 (1992), pp. 9–10.

PC-PROG

Areas covered by the software

Linear programming, mixed-integer linear programming, quadratic programming

Hardware/software environment

Designed specifically for an IBM-compatible PC environment, PC-PROG requires 512KB RAM. It runs interactively and has a full-screen editor to assist in model entry. Windows are available for displaying algorithm progress, viewing the results, and providing on-line help. The system is available in a number of versions that have different limits on problem sizes. These include a Personal version (50 variables, 50 constraints, with at most 1000 nonzeros in the constraint matrix), a Midsize version (250 variables, 250 constraints, 1500 nonzeros), and a Professional version (1000 variables, 1000 constraints, 8000 nonzeros).

Contact address

Hugo Uyttenhove
Computing and Systems Consultants
8205 Old Deer Trail
Raleigh, NC 27615
Phone: (919) 847-9997
Fax: (919) 848-4657

PITCON

Areas covered by the software

Systems of n nonlinear equations involving an n-dimensional state variable and one scalar parameter variable

Basic algorithms

PITCON implements a continuation algorithm with an adaptive choice of a local coordinate system. At each computed point one of the variables is chosen as the local coordinate direction for the one-dimensional solution manifold of the problem. From a predicted point along the tangent direction, a chord Newton process is applied to obtain the next point. Facilities are incorporated to determine the presence of target points where a specified variable has a given value, of turning points with respect to any variable, or of simple bifurcation points. Algorithms are included for the explicit computation of target or turning points.

PITCON requires a user subroutine for the evaluation of the function and incorporates algorithms for computing finite-difference approximations of the Jacobian, unless a subroutine for the direct computation of the Jacobian is provided. The routines to solve the bordered banded linear systems can be easily exchanged for user-supplied codes. Several different linear solvers are provided with PITCON.

Hardware/software environment

The software is written in standard ANSI Fortran. Single- and double-precision versions of the software are available. Machine dependencies are restricted to a single subroutine.

Contact address

The software can be obtained from *netlib* (`send index from contin`).

References

W. C. Rheinboldt, *Numerical Analysis of Parametrized Nonlinear Equations*, John Wiley & Sons, Inc., New York, 1985.

W. C. Rheinboldt and J. Burkardt, *Algorithm 596: A program for a locally parametrized continuation process*, ACM Trans. Math. Software, 9 (1983), pp. 236–241.

———, *A locally parametrized continuation process*, ACM Trans. Math. Software, 9 (1983), pp. 215–235.

W. C. Rheinboldt and R. Melhem, *A comparison of methods for determining turning points of nonlinear equations*, Computing, 29 (1982), pp. 201–226.

PORT 3

Areas covered by the software

General minimization, nonlinear least squares, separable nonlinear least squares, linear inequalities, linear programming, and quadratic programming. The nonlinear optimizers have unconstrained and bound-constrained variants.

Basic algorithms

The nonlinear optimizers use trust-region algorithms. Gradients and Jacobians can be provided by the caller or approximated automatically by finite differences. The general minimization routines use either a quasi-Newton approximation to the Hessian matrix or a Hessian provided by the caller; the nonlinear least squares routines adaptively switch between the Gauss–Newton Hessian approximation and an "augmented" approximation that uses a quasi-Newton update. Function and, if necessary, gradient values may be provided either by subroutines or by reverse communication.

There is a special separable nonlinear least squares solver for the case of one nonlinear variable; it uses Brent's one-dimensional minimization algorithm for the nonlinear variable. Brent's algorithm is also available by itself, as is an implementation of the Nelder–Mead simplex method.

The feasible point (linear inequalities) and linear and quadratic programming routines start by taking steps through the interior and end with an active set strategy. The quadratic programming solvers use the Bunch–Kaufman factorization and thus can find local minimizers of indefinite problems.

None of the solvers is meant for large numbers of variables. When there are n variables and m equations (where $m = 1$ for general minimization), the nonlinear solvers require $O(n^2 m)$ or $O(n^3)$ arithmetic operations per iteration. The linear and quadratic solvers use dense-matrix techniques.

Hardware/software environment

Software is written in ANSI Fortran 77, with single- and double-precision versions of all solvers. Machine-dependent constants are provided by subroutines I1MACH, R1MACH, and D1MACH.

Contact address

> Judith Macor
> AT&T Bell Laboratories, Room 3D-588
> 600 Mountain Avenue
> Murray Hill, NJ 07974-0636
> Phone: (908) 582-7710

Additional comments

A subset of the software is available by email from *netlib* and by ftp from `research.att.com` (log in as `netlib` and give your email address as password). The subset includes the general minimizers [D]MN[FGH][B], the nonlinear least squares solvers [D]N2[FGP][B], and the separable least squares solvers [D]NS[FG][B]. Some invocation examples are also available; for details, ask *netlib* to `send index from port/ex`.

References

J. E. Dennis, Jr., D. M. Gay, and R. E. Welsch, *Algorithm 573. NL2SOL—An adaptive nonlinear least-squares algorithm*, ACM Trans. Math. Software, 7 (1981), pp. 369–383.

P. A. Fox, A. D. Hall, and N. L. Schryer, *The PORT mathematical subroutine library*, ACM Trans. Math. Software, 4 (1978), pp. 104–126.

D. M. Gay, *Algorithm 611. Subroutines for unconstrained minimization using a model/trust-region approach*, ACM Trans. Math. Software, 9 (1983), pp. 503–524.

———, *Usage Summary for Selected Optimization Routines*, Computing Science Tech. Report No. 153, AT&T Bell Laboratories, Murray Hill, NJ, 1990. (Postscript for the last report can be obtained by sending the message `send 153 from research/cstr` to `netlib@research.att.com`.)

PROC NLP (from the SAS/OR package)
Areas covered by the software

Nonlinear minimization or maximization with linear constraints

Basic algorithms

Nonlinear least squares with linear constraints—The Levenberg–Marquardt algorithm (Moré) and a hybrid quasi-Newton algorithm (Fletcher and Xu).

Nonlinear min/maximization with linear constraints—A trust-region algorithm [(Gay) and (Moré and Sorensen)], two different Newton–Raphson algorithms using line search or ridging, quasi-Newton algorithms updating an approximation of the Cholesky decomposition or the inverse of the Hessian, various conjugate gradient algorithms with Powell's automatic restart algorithm (Powell) and Fletcher–Reeves, Polak–Ribiere updates, and a double-dogleg algorithm [(Gay) and (Dennis and Mei)].

Nonlinear min/maximization with simple bounds—Nelder–Mead simplex algorithm.

Hardware/software environment

Software is written in C using double-precision arithmetic and will be available for a wide variety of personal computer, workstation, and mainframe operating systems.

Contact address

SAS Institute Inc.
SAS Campus Drive
Cary, NC 27513
Phone: (919) 677-8000

Additional comments

PROC NLP is part of the SAS/OR (Operations Research) package. The current version of PROC NLP is experimental. When PROC NLP goes into production, various extensions will be implemented. In particular, a special algorithm for optimizing a quadratic function with linear constraints will be offered, nonlinear constraints will be implemented by an augmented Lagrangian approach, and a reduced-gradient version with sparse LU decomposition will be provided for large and sparse systems of linear constraints.

References

R. Fletcher and C. Xu, *Hybrid methods for nonlinear least squares*, IMA J. Numer. Anal., 7 (1987), pp. 371–389.

D. M. Gay, *Algorithm 611. Subroutines for unconstrained minimization using a model/trust-region approach*, ACM Trans. Math. Software, 9 (1983), pp. 503–524.

J. J. Moré, *The Levenberg–Marquardt algorithm: Implementation and theory*, in Numerical Analysis, Dundee 1977, G. A. Watson, ed., Lecture Notes in Mathematics 630, Springer-Verlag, Berlin, 1978, pp. 105–116.

J. J. Moré and D. C. Sorensen, *Computing a trust region step*, SIAM J. Sci. Statist. Comput., 4 (1983), pp. 553–572.

M. J. D. Powell, *Restart procedures for the conjugate gradient method*, Math. Programming, 12 (1977), pp. 241–254.

Q01SUBS

Areas covered by the software

Subroutines are available to solve unconstrained quadratic 0-1 programming for both dense and sparse matrices (including concave quadratic problems with box constraints), the maximum clique problem for dense and sparse graphs, and random test-problem generation for all the above problems

Basic algorithms

Branch and bound with depth-first search and dynamic variable selection. Research reports providing the theoretical basis for the algorithms are available on request.

Hardware/software environment

Software is written in Fortran using integer arithmetic in all cases.

Contact address

>Panos Pardalos
>303 Weil Hall
>Department of Industrial Engineering
>University of Florida
>Gainesville, FL 32611
>pardalos@math.ufl.edu

Additional comments

The code is documented by comment statements. User feedback is much appreciated.

References

P. M. Pardalos, *Construction of test problems in quadratic bivalent programming*, ACM Trans. Math. Software, 17 (1991), pp. 74–87.

P. M. Pardalos and G. P. Rodgers, *A branch and bound algorithm for the maximum clique problem*, Comp. Oper. Res., 19 (1992), pp. 363–375.

——, *Computing aspects of a branch and bound algorithm for quadratic zero-one programming*, Computing, 45 (1990), pp. 131–144.

——, *Parallel branch and bound algorithms for quadratic zero-one programs on the hypercube architecture*, Ann. Oper. Res., 22 (1990), pp. 271–292.

QAPP

Areas covered by the software

Quadratic assignment problems. All global solutions can be obtained

Basic algorithms

This is an exact algorithm based on branch and bound. In addition, the algorithm has been parallelized and implemented on shared-memory machines. Research reports providing computational results and the theoretical basis for the algorithm are available on request.

Hardware/software environment

Software is written in Fortran using integer arithmetic in all cases.

Contact address

> Panos Pardalos
> 303 Weil Hall
> Department of Industrial Engineering
> University of Florida
> Gainesville, FL 32611
> pardalos@math.ufl.edu

Additional comments

User feedback is much appreciated.

References

P. M. Pardalos and J. Crouse, *A parallel algorithm for the quadratic assignment problem*, in Proceedings of the Supercomputing 89 Conference, ACM Press, New York, pp. 351–360.

P. M. Pardalos, K. Murthy, and Y. Li, *Computational experience with parallel algorithms for solving the quadratic assignment problem*, in Computer Science and Operations Research: New Developments in Their Interface, O. Balci, R. Sharda, and S. Zenios, eds., Pergamon Press, Elmsford, NY, 1992, pp. 267–278.

QPOPT

Areas covered by the software

QPOPT is a Fortran package designed to solve linear and quadratic programming problems of the following general form:

$$\operatorname*{minimize}_{x \in \mathbb{R}^n} \quad f(x)$$

$$\text{subject to} \quad \ell \leq \left\{ \begin{array}{c} x \\ Ax \end{array} \right\} \leq u,$$

where A is an $m_L \times n$ matrix (m_L may be zero) and $f(x)$ is one of the following:

FP:	None	(find a feasible point for the constraints)
LP:	$c^T x$	(a linear program)
QP1:	$\frac{1}{2} x^T H x$	H symmetric,
QP2:	$c^T x + \frac{1}{2} x^T H x$	H symmetric,
QP3:	$\frac{1}{2} x^T H^T H x$	H $m \times n$ upper trapezoidal,
QP4:	$c^T x + \frac{1}{2} x^T H^T H x$	H $m \times n$ upper trapezoidal,

with c an n-vector. In QP1 and QP2, there is no restriction on H apart from symmetry. If the quadratic function is convex, a global minimum is found; otherwise, a local minimum is found. The method used is most efficient when many constraints or bounds are active at the solution. If H is positive semi-definite, it is usually more efficient to use the package LSSOL.

Basic algorithms

QPOPT uses a two-phase, active-set method based on an inertia-controlling method that maintains a Cholesky factorization of the reduced Hessian. QPOPT treats all matrices as dense and is not intended for sparse problems.

Hardware/software environment

The QPOPT package contains approximately 14,700 lines of Fortran, of which about 75% are comments. The source code and example program for QPOPT are distributed on a floppy disk. The code is also available via Internet using ftp. A MATLAB interface for QPOPT is also available.

QPOPT includes calls to both Level-1 (vector) and Level-2 (matrix-vector) Basic Linear Algebra Subroutines (BLAS). They may be replaced by versions of the BLAS routines that have been tuned to a particular machine.

QPOPT is written in ANSI Fortran 77 and should run without alteration on any machine with a Fortran 77 compiler. The code was developed on a DECstation 3100 using the MIPS f77 compiler and has been installed on most types of PC, workstation, and mainframe.

Contact addresses

Walter Murray
Department of Operations Research
Terman Engineering Center
Stanford University
Stanford, CA 94305-4022

Philip E. Gill
Department of Mathematics
University of California, San Diego
9500 Gilman Drive
La Jolla, CA 92093-0112

Additional comments

QPOPT is essentially identical to the routine E04NFF of the NAG Fortran Library. E04NFF was introduced at Mark 16.

The method of QPOPT is similar to the method of QPSOL, which was distributed by Stanford University between 1983 and 1991. However, QPOPT is a substantial improvement over QPSOL in both functionality and reliability.

References

P. E. Gill and W. Murray, *Numerically stable methods for quadratic programming*, Math. Programming, 14 (1978), pp. 349–372.

P. E. Gill, W. Murray, M. A. Saunders, and M. H. Wright, *Inertia-controlling methods for general quadratic programming*, SIAM Rev., 33 (1991), pp. 1–36.

NAG Fortran Library Manual, Mark 16, NAG Ltd., Oxford, England, UK, 1993.

SPEAKEASY

Areas covered by the software

General tool for solving numerical problems, including operations research problems

Basic algorithms

Speakeasy is a toolbox that provides access to many numerical libraries. These libraries include LINPACK, EISPACK, FFTPACK, and some components of NAG. Included in the Speakeasy package is a complete set of matrix operations and algebras for matrices, arrays, sets, and time series. The environment allows for easy formulation of problems in a notation that is familiar to technical users. A complete interactive graphics package is included. A recently available graphical user interface provides menuing capabilities.

For problems in operations research, Speakeasy provides interfaces with selected NAG routines and to IBM's Optimization Subroutine Library.

Hardware/software environment

Available for IBM mainframes under MVS and VM, IBM RISC System/6000, DEC VAX under VMS, DECsystems and DECstations, and Sun workstations.

Contact address

>Speakeasy Computer Corporation
>224 South Michigan Avenue
>Chicago, IL 60604
>Phone: (312) 427-2400

Additional comments

The toolbox is continually upgraded and expanded.

SQP

Areas covered by the software

Nonlinear programming

Basic algorithms

SQP uses an implementation of Powell's successive quadratic programming algorithm and is aimed specifically at large, sparse nonlinear programs. It solves the quadratic programming subproblems by using a sparsity-exploiting reduced-gradient method. Sparse data structures are used for the constraint Jacobian, and there is an option to represent the approximate Hessian as a small set of vectors using a limited-memory updating scheme.

SQP requires the same user-supplied subroutines as GRG2 and has similar subroutine and data file interfaces. The entry describing GRG2 contains more details.

Hardware/software environment

SQP is written in ANSI Fortran. Machine dependencies are relegated to the subroutine INITLZ, which defines three machine-dependent constants.

Contact address

>Leon Lasdon
>MSIS Department
>College of Business Administration
>University of Texas
>Austin, TX 78712-1175
>Phone: (512) 471-9433

References

Y. Fan, S. Sarkar, and L. Lasdon, *Experiments with successive quadratic programming algorithms*, J. Optim. Theory Appl., 56 (1988), pp. 359–383.

D. Mahidhara and L. Lasdon, *An SQP Algorithm for Large Sparse Nonlinear Programs*, Working Paper, Management and Information Systems Department, College of Business Administration, University of Texas, Austin, TX, 1991.

TENMIN

Areas covered by the software

Unconstrained optimization

Basic algorithm

This package is intended for solving unconstrained optimization problems where the number of variables n is not too large ($n < 100$, say), so that the cost of storing one $n \times n$ matrix and factoring it at each iteration is acceptable. The package allows the user to choose between a tensor method for unconstrained optimization and a standard method based on a quadratic model. The tensor method bases each iteration upon a specially constructed fourth-order model of the objective function that is not significantly more expensive to form, store, or solve than the standard quadratic model. Both methods calculate the Hessian matrix and gradient vector, either analytically or by finite differences, at each iteration. The step-selection strategy is a line search. The tensor method requires significantly fewer iterations and function evaluations to solve most unconstrained optimization problems than the standard method, and also solves a somewhat wider range of problems. It is especially useful for problems in which the Hessian matrix at the solution is singular.

The software can be called with an interface where the user supplies only the function, number of variables, and starting point; default choices are made for all other input parameters. An alternative interface allows the user to specify any input parameters that are different from the defaults.

Hardware/software environment

The software is written in Fortran 77 using double precision. The only machine dependency is upon machine epsilon, which can be either calculated by the software or provided by the user.

Contact address

> Robert B. Schnabel
> Department of Computer Science
> University of Colorado
> Boulder, CO 80309-0430
> Phone: (303) 492-7554
> bobby@cs.colorado.edu.

References

T. Chow, E. Eskow, and R. B. Schnabel, *A Software Package for Unconstrained Optimization Using Tensor Methods*, Tech. Report CU-CS-492-90, Department of Computer Science, University of Colorado, Boulder, CO, December 1990.

R. B. Schnabel and T. Chow, *Tensor methods for unconstrained optimization using second derivatives*, SIAM J. Optim., 1 (1991), pp. 293–315.

TENSOLVE

Areas covered by the software
Nonlinear equations and nonlinear least squares

Basic algorithms
This package find roots of systems of n nonlinear equations in n unknowns, or minimizers of the sum of squares of $m > n$ nonlinear equations in n unknowns. It allows the user to choose between a tensor method based on a quadratic model and a standard method based on a linear model. Both models calculate an analytic or finite-difference Jacobian matrix at each iteration. The tensor method augments the linear model with a low-rank, second-order term that is chosen so that the model is hardly more expensive to form, store, or solve than the standard linear model. Either a line-search or trust-region step-selection strategy may be selected. The tensor method is significantly more efficient than the standard method in iterations, function evaluations, and time. It is especially useful on problems where the Jacobian matrix at the solution is singular.

The software can be called with an interface where the user supplies only the function, number of variables, and starting point; default choices are made for all other input parameters. An alternative interface allows the user to specify any input parameters that are different from the defaults.

Hardware/software environment
The software is written in Fortran 77 using double precision. The only machine dependency is upon machine epsilon, which is either calculated by the software or provided by the user.

Contact address
> Robert B. Schnabel
> Department of Computer Science
> University of Colorado
> Boulder, CO 80309-0430
> Phone: (303) 492-7554
> bobby@cs.colorado.edu.

Additional comments
A version of this package that efficiently handles large, sparse problems is available in preliminary form.

Reference
R. B. Schnabel and P. Frank, *Tensor methods for nonlinear equations*, SIAM J. Numer. Anal., 21 (1984), pp. 814–843.

SOFTWARE PACKAGES

TN/TNBC

Areas covered by the software

Unconstrained minimization and minimization subject to bound constraints. The software is especially well suited to problems with large numbers of variables.

Basic algorithms

TN uses a truncated Newton method based on a line search. Truncated Newton methods compute an approximation to the Newton direction by approximately solving the Newton equations using an iterative method. In this software, the conjugate gradient method is used as the iterative solver.

Both easy-to-use and customized versions are provided in both the unconstrained and bound-constrained cases.

Hardware/software environment

Software is written in ANSI Fortran 77. A single machine constant is set automatically. The software is currently available only in double precision.

Contact address

Software can be obtained from *netlib* (`send tn from opt`) or from

> Stephen G. Nash
> ORAS Department
> George Mason University
> Fairfax, VA 22030
> Phone: (703) 993-1678
> snash@gmuvax.gmu.edu

References

S. G. Nash, *Newton-like minimization via the Lanczos method*, SIAM J. Numer. Anal., 21 (1984), pp. 770–788.

——, *User's guide for* TN/TNBC, Tech. Report 397, Department of Mathematical Sciences, The Johns Hopkins University, Baltimore, MD, 1984.

TNPACK

Areas covered by the software

Nonlinear unconstrained minimization of large-scale separable problems

Basic algorithms

A truncated Newton method for unconstrained minimization has been specifically developed for large-scale separable problems. TNPACK uses a preconditioned conjugate gradient method to solve the Newton equations approximately at every step. Modifications are incorporated to handle indefiniteness of both the Hessian and the preconditioner. The preconditioning matrix (usually a sparse approximation to the Hessian) is provided by the user. It is factored by a sparse modified Cholesky factorization based on the Yale Sparse Matrix Package. TNPACK is intended to solve complex problems that arise in practical applications, such as computational chemistry and biology, where a natural separability or hierarchy in complexity exists among the different functional components. The user can adapt details of the algorithm to suit the problem at hand (for example, by preconditioning and variable reordering).

Hardware/software environment
Software is written in double-precision ANSI Fortran 77.

Contact address
The package will soon be available as part of the ACM Transactions on Mathematical Software collection. Interested users can also contact

>Tamar Schlick
>Courant Institute of Mathematical Sciences
>251 Mercer Street
>New York, NY 10012
>Phone: (212) 998-3116
>schlick@acfclu.nyu.edu

References

T. Schlick and A. Fogelson, *TNPACK—A truncated Newton minimization package for large-scale problems. I: Algorithm and usage*, ACM. Trans. Math. Software, 18 (1992), pp. 46–70.

T. Schlick and A. Fogelson, *TNPACK—A truncated Newton minimization package for large-scale problems. II: Implementation examples*, ACM. Trans. Math. Software, 18 (1992), pp. 71–111.

T. Schlick and W. K. Olson, *Supercoiled DNA structure and dynamics by computer simulations*, J. Molecular Biol., 223 (1992), pp. 1089–1119.

T. Schlick and M. L. Overton, *A powerful truncated Newton method for potential energy functions*, J. Comput. Chem., 8 (1987), pp. 1025–1039.

UNCMIN
Areas covered by the software
Unconstrained optimization

Basic algorithms

UNCMIN is a modular package based on a Newton or quasi-Newton approach. It allows the user to select from various options for calculating or approximating derivatives and for the step-selection strategy. The Hessian matrix may be calculated either analytically or by finite differences (Newton's method) or the BFGS update (quasi-Newton). The gradient vector may be supplied by the user or calculated by finite differences. Options for step selection include line search, double dogleg trust region, and a hookstep trust-region method. Any combination of these options is permitted.

The software can be called with an interface where the user supplies only the function, number of variables, and starting point; default choices are made for all other input parameters. In this case the method is a BFGS method with line-search and finite-difference gradients. An alternative interface allows the user to specify any input parameters that are different from the defaults.

The package comes with an extensive user's manual and test problems together with their data. It also includes a second version where the function is supplied via reverse communication.

Hardware/software environment

The software is written in standard Fortran. It is provided in single precision, but simple instructions for conversion to double precision are included. The only machine dependency is upon machine precision, which is either calculated by the software or provided by the user.

Contact address

> Robert B. Schnabel
> Department of Computer Science
> University of Colorado
> Boulder, CO 80309-0430
> Phone: (303) 492-7554
> bobby@cs.colorado.edu

Additional comments

The software corresponds closely, although not exactly, to the modular set of algorithms in the appendix of the book by Dennis and Schnabel.

References

J. E. Dennis Jr. and R. B. Schnabel, *Numerical Methods for Unconstrained Optimization and Nonlinear Equations*, Prentice Hall, Englewood Cliffs, NJ, 1983.

R. B. Schnabel, J. E. Koontz, and B. E. Weiss, *A modular system of algorithms for unconstrained minimization*, ACM Trans. Math. Software, 11(1985), pp. 419–440.

VE08

Areas covered by the software

Bound-constrained nonlinear optimization with an emphasis on large-scale problems

Basic algorithms

VE08 is a line-search method with a search direction obtained by a truncated conjugate gradient technique. The bounds are handled by bending the search direction on the boundary of the feasible domain. VE08 also contains provision for estimating gradients by finite difference, if they are unavailable, or to check the analytic gradients otherwise. It features both Newton and quasi-Newton algorithms.

VE08 exploits the partially separable structure of many large-scale problems to obtain good efficiency. In particular, it uses the partitioned updating technique when a quasi-Newton method is chosen.

Hardware/software environment

VE08 is a standard ANSI Fortran subroutine in double precision. Machine dependencies are restricted to arithmetic constants, which can easily be modified by the user.

Contact address

Software can be obtained from *netlib* (`send ve08 from opt`) or from

>Philippe Toint
>Department of Mathematics
>FUNDP
>61 rue de Bruxelles
>B5000 Namur, Belgium
>pht@math.fundp.ac.be

References

A. Griewank and Ph. L. Toint, *Numerical experiments with partially separable optimization problems*, in Numerical Analysis: Proceedings Dundee 1983, D. F. Griffiths, ed., Lecture Notes in Mathematics 1066, Springer-Verlag, Berlin, 1984, pp. 203–220.

———, *Partitioned variable metric updates for large structured optimization problems*, Numer. Math., 39 (1982), pp. 429–448.

Ph. L. Toint, *User's Guide to the Routine VE08 for Solving Partially Separable Bounded Optimization Problems*, Tech. Report 83/1, FUNDP, Namur, Belgium, 1983.

SOFTWARE PACKAGES

VE10

Areas covered by the software

Bound-constrained nonlinear least squares with an emphasis on large-scale problems

Basic algorithms

VE10 is a line-search method in which the search direction is obtained by a truncated conjugate gradient technique. The bounds are handled by bending the search direction on the boundary of the feasible domain. VE10 also contains a provision for estimating gradients by finite difference, if they are unavailable, or to check the analytic gradients otherwise. It features both Newton and quasi-Newton algorithms.

In addition to using the least-squares nature of the problem, VE10 exploits the partially separable structure of many large-scale problems to obtain good efficiency. In particular, it uses a Brown–Dennis partitioned updating technique as the quasi-Newton option for partially separable problems.

Hardware/software environment

VE10 is a standard ANSI Fortran subroutine in double precision. Machine dependencies are restricted to arithmetic constants, which can easily be modified by the user.

Contact address

> Philippe Toint
> Department of Mathematics
> FUNDP
> 61 rue de Bruxelles
> B5000 Namur, Belgium
> pht@math.fundp.ac.be

Additional comments

VE10 can also be obtained from the Harwell Subroutine Library, Harwell Laboratory, Didcot, Oxfordshire, England, UK.

References

Ph. L. Toint, *On large-scale nonlinear least squares calculations*, SIAM J. Sci. Statist. Comput., 8 (1987), pp. 416–435.

———, *VE10AD: A routine for large-scale nonlinear least squares*, Harwell Subroutine Library, Oxfordshire, England, UK, 1987.

VIG and VIMDA

Areas covered by the software

A visual, dynamic, and interactive decision support system for multiple-criteria decision making (VIG) and a visual multiple-criteria decision support system for discrete alternatives with numerical data (VIMDA)

Basic algorithms

Problems are formulated as linear programs with multiple objective functions. Different linear criteria are weighted dynamically with the help of input from the user. The visual interface is described as a *Pareto race.*

Hardware/software environment

VIG is designed for IBM-compatible PC/DOS; 256KB of RAM is needed, and a color graphics monitor is recommended. VIG can handle linear programming constraint matrices with 96 columns and 100 rows, of which 10 rows may constitute the objective functions.

VIMDA runs on IBM-compatible PC machines under DOS and requires 256KB of RAM and a graphics card. A color graphics monitor is strongly recommended.

Codes are distributed via 3.5-inch or 5.25-inch disks.

Contact address

>NumPlan
>P. O. Box 128
>SF–03101 Nummela
>Finland
>Phone: +358 (0) 2271-900

Reference

A Multiple Criteria Decision Support System in Corporate Planning, NumPlan, Nummela, Finland, September 1989.

What's *Best!*

Areas covered by the software

Linear programming, mixed-integer linear programming

Basic algorithms

What's *Best!* provides a spreadsheet interface to the LINDO optimizer.

SOFTWARE PACKAGES

Hardware/software environment

What's *Best!* is available in a number of formats that have different platform requirements. These are summarized in the following table.

	Personal	Commercial	Professional	Industrial	Extended
Variables	400	1,500	4,000	16,000	32,000
Constraints	200	750	2,000	8,000	16,000
Nonzeros	4,000	24,000	32,000	128,000	256,000
Memory[a]	256KB	384KB	1MB	4MB	6MB
Memory[b]	2MB	2MB	2MB	2-4MB	8MB

[a]PC; [b]Macintosh.

What's *Best!* supports the following spreadsheets: Quattro Pro 2.0 and 3.0, Lotus 1-2-3 Release 2.X and 3.X, Excel for the Macintosh, and Symphony.

Contact address

LINDO Systems, Inc.
P. O. Box 148231
Chicago, IL 60614
Phone: (800) 441-BEST and (312) 871-2524
Fax: (312) 871-1777

Appendix: Internet Software

Much of the software described in this book is in the public domain and can be obtained through Internet. There are two main modes of access: electronic mail and ftp.

The *netlib* software library can be accessed by electronic mail. This library includes software written by academic researchers and software that is described in the journal *ACM Transactions on Mathematical Software*.

A long message containing an introduction to *netlib*, a list of contents, and instructions on how to request software can be obtained by sending the message `send index` to the Internet address `netlib@ornl.gov` or `netlib@research.att.com`. For European users, it may be more convenient to access the duplicate collection in Norway, for which the address differs according to the network of origin:

Internet	`netlib@nac.no`
EARN/BITNET	`netlib%nac.no@norunix.bitnet`
X.400	`s=netlib; o=nac; c=no;`
EUNET/uucp	`nac!netlib`

Once the user has identified the appropriate section of *netlib*, additional information can be obtained by requesting the index for that section. For example, the message `send index for toms` generates a return message with a listing of the software available from *ACM Transactions on Mathematical Software*.

Some public domain software can also be obtained from anonymous ftp sites. Given the address of the host machine at the remote ftp site, access is obtained by issuing the `ftp` command followed by the remote host address. The host typically requests a user name, to which the user should respond by typing `anonymous`. This is usually followed by a request for a password; the conventional (though not compulsory) response is to provide the user's electronic mail address.

Once users have obtained access to an anonymous ftp site, they typically need to search the directory hierarchy for the appropriate files. Information can often be found in files with names such as `readme`. The most useful ftp commands are

cd remote-directory: Changes the directory being examined on the remote machine to **remote-directory**.

ls: Lists the files on the current directory of the remote machine.

get remote-file: Downloads the file **remote-file** to the working directory on the local machine.

quit : Disconnects from the remote machine and exits ftp.

A listing of valid commands can be obtained by typing **help**. Additional information on these commands can be found in a manual.

Software from *netlib* can be obtained by anonymous ftp to the hosts **research.att.com** (at AT&T Bell Labs) and **ftp.cs.uow.edu.au** (at the University of Wollongong, Australia). If the software files are long or if many files are needed, ftp is usually a more convenient form of access to *netlib* than electronic mail. The **ornl.gov** site does not currently support anonymous ftp.

We show below a sample **ftp** session with **dimacs.rutgers.edu**. This site contains the DIMACS collection at Rutgers University, which is described under the NETFLOW listing. In this session we download only the file **readme**, which contains information on the directory structure for **pub/netflow**. We have omitted much of the output provided by ftp.

```
ftp dimacs.rutgers.edu
Connected to dimacs.rutgers.edu.
220 dimacs.rutgers.edu FTP server
Name (dimacs.rutgers.edu:more): anonymous
331 Guest login ok, send e-mail address as password.
Password: user's e-mail address
230 Guest login ok, access restrictions apply.
ftp> cd pub/netflow
250 CWD command successful.
ftp> get readme
ftp> quit
221 Goodbye.
```

In a typical session we would use the information in the **readme** file to explore the directory structure and obtain the addresses of other files.

Additional information on the Internet can be found in *The Whole Internet User's Guide & Catalog* by E. Kroll, O'Reilly & Associates, Sebastopol, CA, 1992.

References

[1] R. K. Ahuya, T. L. Magnanti, and J. B. Orlin, *Network Flows: Theory, Algorithms, and Applications*, Prentice Hall, Englewood Cliffs, NJ, 1993.

[2] E. L. Allgower and K. Georg, *Numerical Continuation: An Introduction*, Springer-Verlag, Berlin, 1990.

[3] D. M. Bates and D. G. Watts, *Nonlinear Regression Analysis and Its Applications*, John Wiley & Sons, Inc., New York, 1988.

[4] D. P. Bertsekas, *Constrained Optimization and Lagrange Multiplier Methods*, Academic Press, New York, 1982.

[5] ———, *Linear Network Optimization: Algorithms and Codes*, MIT Press, Cambridge, MA, 1991.

[6] ———, *Auction algorithms for network flow problems: A tutorial introduction*, Comput. Optim. Appl., 1 (1992), pp. 7–66.

[7] A. Björck, *Least squares methods*, in Handbook of Numerical Analysis, P. G. Ciarlet and J. L. Lions, eds., North–Holland, Amsterdam, 1990, pp. 465–647.

[8] V. Chvátal, *Linear Programming*, W. H. Freeman and Company, New York, 1983.

[9] T. F. Coleman, *Large scale numerical optimization: Introduction and overview*, in Encyclopedia of Computer Science and Technology, J. Williams and A. Kent, eds., Marcel Dekker, New York, 1993, pp. 167–195.

[10] A. R. Conn, N. I. M. Gould, and Ph. L. Toint, *LANCELOT*, Springer Series in Computational Mathematics, Springer-Verlag, Berlin, 1992.

[11] ———, *Large-scale nonlinear constrained optimization*, in Proceedings of the Second International Conference on Industrial and Applied Mathematics, R. E. O'Malley, ed., Society for Industrial and Applied Mathematics, Philadelphia, PA, 1992, pp. 51–70.

[12] R. S. Dembo, J. M. Mulvey, and S. A. Zenios, *Large-scale nonlinear network models and applications*, Oper. Res., 37 (1989), pp. 353–372.

[13] J. E. Dennis and R. B. Schnabel, *Numerical Methods for Unconstrained Optimization and Nonlinear Equations*, Prentice Hall, Englewood Cliffs, NJ, 1983.

[14] ———, *A view of unconstrained optimization*, in Optimization, G. L. Nemhauser, A. H. G. Rinnooy Kan, and M. J. Todd, eds., North–Holland, Amsterdam, 1989, pp. 1–72.

[15] R. Fletcher, *Practical Methods of Optimization*, 2nd ed., John Wiley & Sons, Inc., New York, 1987.

[16] M. Florian, *Mathematical programming applications in national, regional, and urban planning*, in Mathematical Programming—Recent Developments and Applications, M. Iri and K. Tanabe, eds., Kluwer–Nijhoff, Boston, 1989.

[17] P. E. Gill, W. Murray, M. A. Saunders, and M. H. Wright, *Constrained nonlinear programming*, in Optimization, G. L. Nemhauser, A. H. G. Rinnooy Kan, and M. J. Todd, eds., North–Holland, Amsterdam, 1989, pp. 171–210.

[18] P. E. Gill, W. Murray, and M. H. Wright, *Practical Optimization*, Academic Press, New York, 1981.

[19] ———, *Numerical Linear Algebra and Optimization*, Addison–Wesley, Reading, MA, 1991.

[20] G. H. Golub and C. F. Van Loan, *Matrix Computations*, 2nd ed., The Johns Hopkins University Press, Baltimore, MD, 1989.

[21] C. C. Gonzaga, *Path-following methods for linear programming*, SIAM Rev., 34 (1992), pp. 167–224.

[22] W. W. Hager, R. Horst, and P. M. Pardalos, *Mathematical programming—A computational perspective*, in Handbook of Statistics, C. R. Rao, ed., Vol. 9, Elsevier, New York, 1993, pp. 201–278.

[23] N. Karmarkar, *A new polynomial-time algorithm for linear programming*, Combinatorica, 4 (1984), pp. 373–395.

[24] J. L. Kennington and R. V. Helgason, *Algorithms for Network Programming*, John Wiley & Sons, Inc., New York, 1980.

[25] C. L. Lawson and R. J. Hanson, *Solving Least Squares Problems*, Prentice Hall, Englewood Cliffs, NJ, 1974.

[26] I. J. Lustig, R. E. Marsten, and D. F. Shanno, *Computational experience with a primal-dual interior point method for linear programming*, Linear Algebra Appl., 152 (1991), pp. 191–222.

[27] J. J. Moré and D. C. Sorensen, *Newton's method*, in Studies in Numerical Analysis, G. H. Golub, ed., Mathematical Association of America, Washington, D.C., 1984, pp. 29–82.

[28] B. A. Murtagh, *Advanced Linear Programming: Computation and Practice*, McGraw–Hill, New York, 1981.

[29] G. L. Nemhauser and L. A. Wolsey, *Integer and Combinatorial Optimization*, John Wiley & Sons, Inc., New York, 1988.

[30] ———, *Integer programming*, in Optimization, G. L. Nemhauser, A. H. G. Rinnooy Kan, and M. J. Todd, eds., North–Holland, Amsterdam, 1989, pp. 447–527.

[31] J. Nocedal, *Theory of algorithms for unconstrained optimization*, Acta Numerica 1992, Cambridge University Press, Cambridge, pp. 199–242.

[32] W. C. Rheinboldt, *Numerical Analysis of Parametrized Nonlinear Equations*, John Wiley & Sons, Inc., New York, 1986.

[33] A. H. G. Rinnooy Kan and G. T. Timmer, *Global optimization*, in Optimization, G. L. Nemhauser, A. H. G. Rinnooy Kan, and M. J. Todd, eds., North–Holland, Amsterdam, 1989, pp. 631–662.

[34] A. Schrijver, *Theory of Linear and Integer Programming*, John Wiley & Sons, Inc., New York, 1986.

[35] G. A. F. Seber and C. J. Wild, *Nonlinear Regression*, John Wiley & Sons, Inc., New York, 1989.

[36] L. T. Watson, *Numerical linear algebra aspects of globally convergent homotopy methods*, SIAM Rev., 28 (1986), pp. 529–545.